"One of the year's ten best books."

—PEOPLE

"A riveting nonfiction thriller . . . swashbuckling reading."

—NEW YORK TIMES

"No movie will match the real-life horror described in Richard Preston's *The Hot Zone*."

—TIME

"A bone-chilling account of a close encounter with a lethal virus . . . a totally convincing page-turner, proving that truth is scarier than fiction."

—KIRKUS REVIEWS

"Mesmerizing."

"This work of nonfiction is more terrifying than any sci-fi nightmare."

"Horrifying and riveting . . . Preston exposes a real-life nightmare potentially as lethal as the fictive runaway germs in Michael Crichton's *The Andromeda Strain*."

"Utterly engrossing . . . will make your blood curdle."

"As spine-chilling a narrative as I've read in recent years . . . vivid and cinematic."

"Fascinating—and frightening."

BY RICHARD PRESTON

First Light
American Steel
The Hot Zone
The Cobra Event

THE HOT ZONE

RICHARD PRESTON

ANCHOR BOOKS
A DIVISION OF RANDOM HOUSE, INC.
NEW YORK

FIRST ANCHOR BOOKS EDITION, AUGUST 1995

Portions of this work were originally published in *The New Yorker.*

GRATEFUL ACKNOWLEDGMENT IS MADE TO THE FOLLOWING FOR
PERMISSION TO USE BOTH PUBLISHED AND UNPUBLISHED MATERIAL:
DAN W. DALGARD: Brief excerpts from "Chronology of Events" by Dan W.
Dalgard. Copyright © 1989, 1994 by Dan W. Dalgard. Reprinted by permission.
KARL M. JOHNSON: Excerpt from a letter by Karl M. Johnson to Richard Preston.
Reprinted by permission.
THE WASHINGTON POST: Excerpt from "Deadly Ebola Virus Found in Va.
Laboratory" by D'Vera Cohn (12/1/89). Copyright © 1989 by The Washington
Post. Reprinted by permission.

Library of Congress Cataloging-in-Publication Data

Preston, Richard, 1954–
The hot zone / Richard Preston. — 1st Anchor Books ed.
p.   cm.
Originally published: New York: Random House, 1994.
1. Ebola virus disease—Virginia—Reston.   2. Ebola virus disease—Africa.
3. Primates as laboratory animals.   I. Title.
RC140.5.P74   1995b
614.5′7—dc20   95-5751
CIP

ISBN 0-385-49522-6

www.anchorbooks.com

Printed in the United States of America

30 29 28 27 26 25 24 23 22

TO FREDERIC DELANO GRANT, JR.,
ADMIRED BY ALL WHO KNOW HIM

THE AUTHOR GRATEFULLY ACKNOWLEDGES A
RESEARCH GRANT FROM
THE ALFRED P. SLOAN FOUNDATION

This book describes events between 1967 and 1993. The incubation period of the viruses in this book is less than twenty-four days. No one who suffered from any of the viruses or who was in contact with anyone suffering from them can catch or spread the viruses outside of the incubation period. None of the living people referred to in this book suffer from a contagious disease. The viruses cannot survive independently for more than ten days unless the viruses are preserved and frozen with special procedures and laboratory equipment. Thus none of the locations in Reston or the Washington, D.C., area described in this book is infective or dangerous.

The second angel poured his bowl into the sea, and it became like the blood of a dead man.

—APOCALYPSE

To the Reader

This book is nonfiction. The story is true, and the people are real. I have occasionally changed the names of characters, including "Charles Monet" and "Peter Cardinal." When I have changed a name, I state so in the text.

The dialogue comes from the recollections of the participants, and has been extensively cross-checked. At certain moments in the story, I describe the stream of a person's thoughts. In such instances, I am basing my narrative on interviews with the subjects in which they have recalled their thoughts often repeatedly, followed by fact-checking sessions in which the subjects confirmed their recollections. If you ask a person, "What were you thinking?" you may get an answer that is richer and more revealing of the human condition than any stream of thoughts a novelist could invent. I try to see through people's faces into their minds and listen through their words into their lives, and what I find there is beyond imagining.

—RICHARD PRESTON

Contents

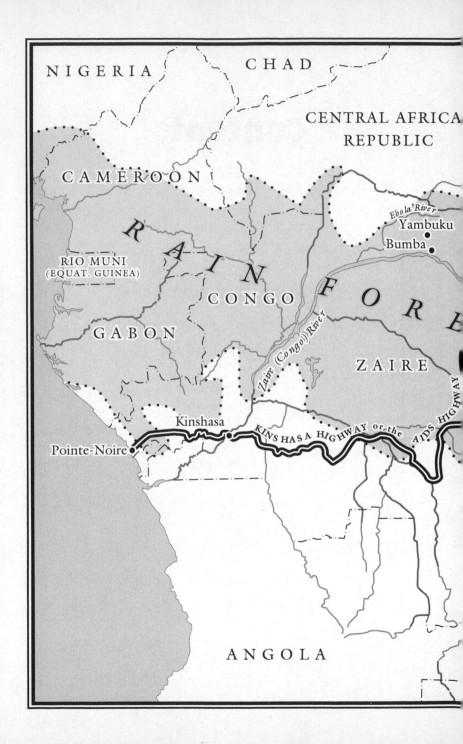

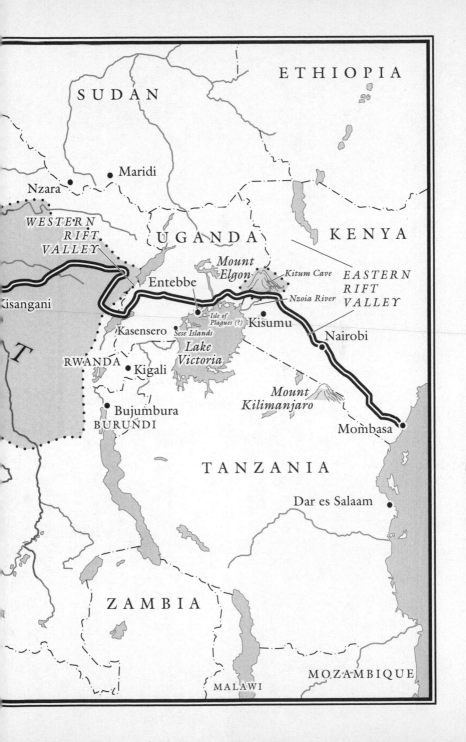

INFECTIOUS AREA
NO UNAUTHORIZED
ENTRY

**TO OPEN THIS DOOR,
PLACE ID CARD ON SENSOR.**

LOCK　　　　　　　　　　**UNLOCK**

PROCESSING . . .
YOU ARE CLEARED TO ENTER. . . .

SUITE AA-5

INVESTIGATOR:
COL. NANCY JAAX

AGENTS IN USE:
UNKNOWN

PROCEED FORWARD.

BIOSAFETY LEVEL

LOCKER ROOM

STATUS:
FEMALE

REMOVE EVERYTHING TOUCHING THE SKIN: CLOTHING, RINGS, CONTACT LENSES, ETC. CHANGE INTO STERILE SURGICAL SCRUBS.

YOU ARE CLEARED TO ENTER. . . .

BIOSAFETY LEVEL

CAUTION:
ULTRAVIOLET LIGHT

BIOSAFETY LEVEL

STAGING AREA

ALARMS:
ENABLED

SPACE-SUIT STATUS:
READY

CAUTION

BIOHAZARD

BIOSAFETY LEVEL

AIR-LOCK DOOR/DECON SHOWER

DO NOT ENTER WITHOUT WEARING
SPACE SUIT

ID CODE PLEASE?
YOU ARE CLEARED TO ENTER. . . .

PART ONE

THE SHADOW
OF MOUNT ELGON

Something
in the Forest

Charles Monet was a loner. He was a Frenchman who lived by himself in a little wooden bungalow on the private lands of the Nzoia Sugar Factory, a plantation in western Kenya that spread along the Nzoia River within sight of Mount Elgon, a huge, solitary, extinct volcano that rises to a height of fourteen thousand feet near the edge of the Rift Valley. Monet's history is a little obscure. As with so many expatriates who end up in Africa, it is not clear what brought him there. Perhaps he had been in some kind of trouble in France, or perhaps he had been drawn to Kenya by the beauty of the country. He was an amateur naturalist, fond of birds and animals but not of humanity in general. He was fifty-six years old, of medium height and medium build, with smooth, straight brown hair; a good-looking man. It seems that his only close friends were women who lived in towns around the mountain, yet even they could not recall much about him for the doctors who investigated his death. His job was to take care of the sugar factory's water-pumping machinery, which drew water from the Nzoia River and delivered it to many miles of sugar-cane fields. They say that he spent most of his day inside the pump house by the river, as if it pleased him to watch and listen to machines doing their work.

So often in a case like this, it's hard to pin down the details. The doctors remember the clinical signs, because no one who has seen the effects of a Biosafety Level 4 hot agent on a human being can ever forget them, but the effects pile up, one after the other, until they obliterate the person beneath them. The case of Charles Monet emerges in a cold geometry of clinical fact mixed with flashes of horror so brilliant and disturbing that we draw back and blink, as if we are staring into a discolored alien sun.

Monet came into the country in the summer of 1979, around the time that the human immunodeficiency virus, or HIV, which causes AIDS, made a final breakout from the rain forests of central Africa and began its long burn through the human race. AIDS had already fallen like a shadow over the population, although no one yet knew it existed. It had been spreading quietly along the Kinshasa Highway, a transcontinental road that wanders across Africa from east to west and passes along the shores of Lake Victoria within sight of Mount Elgon. HIV is a highly lethal but not very infective Biosafety Level 2 agent. It does not travel easily from person to person, and it does not travel through the air. You don't need to wear a biohazard space suit while handling blood infected with HIV.

Monet worked hard in the pump house during the week, and on his weekends and holidays he would visit forested areas near the sugar factory. He would bring food with him, and he would scatter it around and watch while birds and animals ate it. He could sit in perfect stillness while he observed an animal. People who knew him recalled that he was affectionate with wild monkeys, that he had a special way with them. They said that he would sit holding a piece of food while a monkey approached him, and the animal would eat from his hand.

In the evenings, he kept to himself in his bungalow. He

had a housekeeper, a woman named Johnnie, who cleaned up and prepared his meals. He was teaching himself how to identify African birds. A colony of weaverbirds lived in a tree near his house, and he spent time watching them build and maintain their baglike nests. They say that one day near Christmas he carried a sick bird into his house, where it died, perhaps in his hands. The bird may have been a weaverbird—no one knows—and it may have died of a Level 4 virus—no one knows. He also had a friendship with a crow. It was a pied crow, a black-and-white bird that people in Africa sometimes make into a pet. This crow was a friendly, intelligent bird that liked to perch on the roof of Monet's bungalow and watch his comings and goings. When the crow was hungry, it would land on the veranda and walk indoors, and Monet would feed it scraps of food from his table.

He walked to work every morning through the cane fields, a journey of two miles. That Christmas season, the workers had been burning the fields, and so the fields were scorched and black. To the north across the charred landscape, twenty-five miles away, he could see Mount Elgon. The mountain displayed a constantly changing face of weather and shadow, rain and sun, a spectacle of African light. At dawn, Mount Elgon appeared as a slumped pile of gray ridges receding into haze, culminating in a summit with two peaks, which are opposed lips of the eroded cone. As the sun came up, the mountain turned silvery green, the color of the Mount Elgon rain forest, and as the day progressed clouds appeared and hid the mountain from view. Late in the afternoon, near sunset, the clouds thickened and boiled up into an anvil thunderhead that flickered with silent lightning. The bottom of the cloud was the color of charcoal, and the top of the cloud feathered out against the upper air and glowed a dull orange, illuminated by the setting sun, and above the cloud the sky was deep blue and gleamed with a few tropical stars.

He had a number of women friends who lived in the town of Eldoret, to the southeast of the mountain, where the people are poor and live in shacks made of boards and metal. He gave money to his women friends, and they, in return, were happy to love him. When his Christmas vacation arrived, he formed a plan to go camping on Mount Elgon, and he invited one of the women from Eldoret to accompany him. No one seems to remember her name.

Monet and his friend drove in a Land Rover up the long, straight red-dirt road that leads to Endebess Bluff, a prominent cliff on the eastern side of the volcano. The road was volcanic dust, as red as dried blood. They climbed onto the lower skirts of the volcano and went through cornfields and coffee plantations, which gave way to grazing land, and the road passed old, half-ruined English colonial farms hidden behind lines of blue-gum trees. The air grew cool as they went higher, and crested eagles flapped out of cedar trees. Not many tourists visit Mount Elgon, so Monet and his friend were probably driving the only vehicle on the road, although there would have been crowds of people walking on foot, villagers who cultivate small farms on the lower slopes of the mountain. They approached the frayed outer edge of the Mount Elgon rain forest, passing by fingers and islands of trees, and they passed the Mount Elgon Lodge, an English inn built in the earlier part of the century, now falling into disrepair, its walls cracked and its paint peeling off in the sun and rain.

Mount Elgon straddles the border between Uganda and Kenya, and is not far from Sudan. The mountain is a biological island of rain forest in the center of Africa, an isolated world rising above dry plains, fifty miles across, blanketed with trees, bamboo, and alpine moor. It is a knob in the backbone of central Africa. The volcano grew up seven to ten million years ago, producing fierce eruptions and explosions of ash, which repeatedly wiped out the forests that grew on its slopes, until it attained a tremendous height. Before Mount Elgon was eroded down, it may have been the highest mountain in Africa, higher than Kilimanjaro is today. It is still the widest. When the sun rises, it throws the shadow of Mount Elgon westward and deep into Uganda, and when the sun sets, the shadow reaches eastward across Kenya. Within the shadow of Mount Elgon lie villages and cities inhabited by various tribal groups, including the Elgon Masai, a pastoral people who came from the north and settled around the mountain some centuries ago, and who raise cattle. The lower slopes of the mountain are washed with gentle rains, and the air remains cool and fresh all year, and the volcanic soil produces rich crops of corn. The villages form a ring of human settlement around the volcano, and the ring is steadily closing around the forest on its slopes, a noose that is strangling the wild habitat of the mountain. The forest is being cleared away, the trees are being cut down for firewood or to make room for grazing land, and the elephants are vanishing.

A small part of Mount Elgon is a national park. Monet and his friend stopped at the park gate to pay their entrance fees. A monkey or perhaps a baboon—no one seems to remember—used to hang out around the gate, looking for handouts, and Monet enticed the animal to sit on his shoulder by offering it a banana. His friend laughed, but they stayed perfectly still while the animal ate. They drove a short way up the mountain and pitched their tent in a clearing of moist green grass that sloped down to a stream. The stream gurgled out of the rain forest, and it was a strange color, milky with volcanic dust. The grass was kept short by Cape buffalo grazing it, and was spotted with their dung.

The Elgon forest towered around their campsite, a web of gnarled African olive trees hung with moss and creepers and dotted with a black olive that is poisonous to humans. They heard a scuffle of monkeys feeding in the trees, a hum of insects, an occasional low *huh-huh* call of a monkey. These were colobus monkeys, and sometimes one would come down from a tree and scuttle across the meadow near the tent, watching them with alert, intelligent eyes. Flocks of olive pigeons burst from the trees on swift downward slants, flying at terrific speed, which is their strategy to escape from harrier hawks that can dive on them and rip them apart on the wing. There were camphor trees and teaks and African cedars and red stinkwood trees, and here and there a dark green cloud of leaves mushroomed above the forest canopy. These were the crowns of podocarpus trees, or podos, the largest trees in Africa, nearly as large as California sequoias. Thousands of elephants lived on the mountain then, and they could be heard moving through the forest, making cracking sounds as they peeled bark and broke limbs from trees.

In the afternoon, it would have rained, as it usually does on Mount Elgon, and so Monet and his friend would have stayed in their tent, and perhaps they made love while a thunderstorm hammered the canvas. It grew dark; the rain tapered off. They built a fire and cooked a meal. It was New Year's Eve. Perhaps they celebrated, drinking champagne. The clouds would have cleared off in a few hours, as they usually do, and the volcano would have emerged as a black shadow under the Milky Way. Perhaps Monet stood on the grass at the stroke of midnight and looked at the stars—neck bent backward, unsteady on his feet from the champagne.

On New Year's morning, sometime after breakfast—a cold morning, air temperature in the forties, the grass wet and cold—they drove up the mountain along a muddy track and parked in a small valley below Kitum Cave. They bush-whacked up the valley, following elephant trails that mean-dered beside a little stream that ran through stands of olive trees and grassy meadows. They kept an eye out for Cape buffalo, a dangerous animal to encounter in the forest. The cave opened at the head of the valley, and the stream cas-caded over its mouth. The elephant trails joined at the en-trance and headed inside. Monet and his friend spent the whole of New Year's Day there. It probably rained, and so they would have sat in the entrance for hours while the little stream poured down in a veil. Looking across the valley, they watched for elephants, and they saw rock hyraxes—furry animals the size of groundhogs—running up and down the boulders near the mouth of the cave.

Herds of elephants go inside Kitum Cave at night to obtain minerals and salts. On the plains, it is easy for elephants to find salt in hardpans and dry water holes, but in the rain forest salt is a precious thing. The cave is large enough to hold as many as seventy elephants at a time. They spend the night inside the cave, dozing on their feet or mining the rock with their tusks. They pry and gouge rocks off the walls, and chew them to fragments between their teeth, and swallow the broken bits of rock. Elephant dung around the cave is full of crumbled rock.

Monet and his friend had a flashlight, and they walked back into the cave to see where it went. The mouth of the cave is huge—fifty-five yards wide—and it opens out even wider beyond the entrance. They crossed a platform covered with powdery dry elephant dung, their feet kicking up puffs of dust as they advanced. The light grew dim, and the floor of the cave rose upward in a series of shelves coated with green slime. The slime was bat guano, digested vegetable matter that had been excreted by a colony of fruit bats on the ceiling.

Bats whirred out of holes and flicked through their flashlight beams, dodging around their heads, making high-pitched cries. Their flashlights disturbed the bats, and more bats woke up. Hundreds of bat eyes, like red jewels, looked down on them from the ceiling of the cave. Waves of bat sound rippled across the ceiling and echoed back and forth, a dry, squeaky sound, like many small doors being opened on dry hinges. Then they saw the most wonderful thing about Kitum Cave. The cave is a petrified rain forest. Mineralized logs stuck out of the walls and ceiling. They were trunks of rain-forest trees turned to stone—teaks, podo trees, evergreens. An eruption of Mount Elgon about seven million years ago had buried the rain forest in ash, and the logs had

been transformed into opal and chert. The logs were surrounded by crystals, white needles of minerals that had
grown out of the rock. The crystals were as sharp as hypodermic syringes, and they glittered in the beams of the flashlights.

Monet and his friend wandered through the cave, shining
their lights on the petrified rain forest. Did he run his hands
over the stone trees and prick his finger on a crystal? They
found petrified bones sticking out of the ceiling and walls.
Bones of crocodiles, bones of ancient hippos and ancestors
of elephants. There were spiders hanging in webs among the
logs. The spiders were eating moths and insects.

They came to a gentle rise, where the main chamber wid-
ened to more than a hundred yards across—wider than the
length of a football field. They found a crevice and shined
their lights down to the bottom. There was something
strange down there—a mass of gray and brownish material.
It was the mummified corpses of baby elephants. When elephants walk through the cave at night, they navigate by their
sense of touch, probing the floor ahead of them with the tips
of their trunks. The babies sometimes fall into the crevice.

Monet and his friend continued deeper into the cave, descending a slope, until they came to a pillar that seemed to
support the roof. The pillar was scored with hatch marks and
grooves, the marks of elephant tusks. If the elephants continued to dig away at the base of the pillar, it might eventually
collapse, bringing down the roof of Kitum Cave with it. At
the back of the cave, they found another pillar. This one was
broken. Over it hung a velvety mass of bats, which had
fouled the pillar with black guano—a different kind of guano
from the green slime near the mouth of the cave. These bats
were insect eaters, and the guano was an ooze of digested
insects. Did Monet put his hand in the ooze?

Monet's friend dropped out of sight for several years after
that trip to Mount Elgon. Then, unexpectedly, she surfaced
in a bar in Mombasa, where she was working as a prostitute.

A Kenyan doctor who had investigated the Monet case happened to be drinking a beer in the bar, and he struck up an idle conversation with her and mentioned Monet's name. He was stunned when she said, "I know about that. I come from western Kenya. I was the woman with Charles Monet." He didn't believe her, but she told him the story in enough detail that he became convinced she was telling the truth. She vanished after that meeting in the bar, lost in the warrens of Mombasa, and by now she has probably died of AIDS.

Charles Monet returned to his job at the pump house at the sugar factory. He walked to work each day across the burned cane fields, no doubt admiring the view of Mount Elgon, and when the mountain was buried in clouds, perhaps he could still feel its pull, like the gravity of an invisible planet. Meanwhile, something was making copies of itself inside Monet. A life form had acquired Charles Monet as a host, and it was replicating.

The headache begins, typically, on the seventh day after exposure to the agent. On the seventh day after his New Year's visit to Kitum Cave—January 8, 1980—Monet felt a throbbing pain behind his eyeballs. He decided to stay home from work and went to bed in his bungalow. The headache grew worse. His eyeballs ached, and then his temples began to ache, the pain seeming to circle around inside his head. It would not go away with aspirin, and then he got a severe backache. His housekeeper, Johnnie, was still on her Christmas vacation, and he had recently hired a temporary housekeeper. She tried to take care of him, but she really didn't know what to do. Then, on the third day after his headache started, he became nauseated, spiked a fever, and began to vomit. His vomiting grew intense and turned into dry heaves. At the same time, he became strangely passive. His face lost all appearance of life and set itself into an expressionless mask, with the eyeballs fixed, paralytic, and staring. The eyelids were slightly droopy, which gave him a peculiar appearance, as if his eyes were popping out of his head and half-closed at the same time. The eyeballs themselves seemed almost frozen in their sockets, and they turned bright red. The skin of his face turned yellowish, with brilliant starlike red speckles. He began to look like a zombie. His appearance frightened the temporary housekeeper. She didn't understand the transformation in this man. His personality changed. He became sullen, resentful, angry, and his memory seemed to be blown away. He was not delirious. He could answer questions, although he didn't seem to know exactly where he was.

When Monet failed to show up for work, his colleagues began to wonder about him, and eventually they went to his bungalow to see if he was all right. The black-and-white crow sat on the roof and watched them as they went inside. They looked at Monet and decided that he needed to get to a hospital. Since he was very unwell and no longer able to drive a car, one of his co-workers drove him to a private hospital in the city of Kisumu, on the shore of Lake Victoria. The doctors at the hospital examined Monet, and could not come up with any explanation for what had happened to his eyes or his face or his mind. Thinking that he might have some kind of bacterial infection, they gave him injections of antibiotics, but the antibiotics had no effect on his illness.

The doctors thought he should go to Nairobi Hospital, which is the best private hospital in East Africa. The telephone system hardly worked, and it did not seem worth the effort to call any doctors to tell them that he was coming. He could still walk, and he seemed able to travel by himself. He had money; he understood he had to get to Nairobi. They put him in a taxi to the airport, and he boarded a Kenya Airways flight.

A hot virus from the rain forest lives within a twenty-four-hour plane flight from every city on earth. All of the earth's cities are connected by a web of airline routes. The web is a network. Once a virus hits the net, it can shoot anywhere in a day—Paris, Tokyo, New York, Los Angeles, wherever planes fly. Charles Monet and the life form inside him had entered the net.

The plane was a Fokker Friendship with propellers, a commuter aircraft that seats thirty-five people. It started its engines and took off over Lake Victoria, blue and sparkling, dotted with the dugout canoes of fishermen. The Friendship turned and banked eastward, climbing over green hills quilted with tea plantations and small farms. The commuter flights that drone across Africa are often jammed with people, and this flight was probably full. The plane climbed over belts of forest and clusters of round huts and villages with tin roofs. The land suddenly dropped away, going down in shelves and ravines, and changed in color from green to brown. The plane was crossing the Eastern Rift Valley. The passengers looked out the windows at the place where the human species was born. They saw specks of huts clustered inside circles of thornbush, with cattle trails radiating from the huts. The propellers moaned, and the Friendship passed through cloud streets, lines of puffy Rift clouds, and began to bounce and sway. Monet became airsick.

The seats are narrow and jammed together on these commuter airplanes, and you notice everything that is happening inside the cabin. The cabin is tightly closed, and the air recirculates. If there are any smells in the air, you perceive them. You would not have been able to ignore the man who was getting sick. He hunches over in his seat. There is something wrong with him, but you can't tell exactly what is happening.

He is holding an airsickness bag over his mouth. He coughs a deep cough and regurgitates something into the bag. The bag swells up. Perhaps he glances around, and then you see that his lips are smeared with something slippery and red, mixed with black specks, as if he has been chewing coffee grounds. His eyes are the color of rubies, and his face is an expressionless mass of bruises. The red spots, which a few days before had started out as starlike speckles, have expanded and merged into huge, spontaneous purple shadows: his whole head is turning black-and-blue. The muscles of his face droop. The connective tissue in his face is dissolving, and his face appears to hang from the underlying bone, as if the face is detaching itself from the skull. He opens his mouth and gasps into the bag, and the vomiting goes on endlessly. It will not stop, and he keeps bringing up liquid, long after his stomach should have been empty. The airsickness bag fills up to the brim with a substance known as the *vomito negro,* or the black vomit. The black vomit is not really black; it is a speckled liquid of two colors, black and red, a stew of tarry granules mixed with fresh red arterial blood. It is hemorrhage, and it smells like a slaughterhouse. The black vomit is loaded with virus. It is highly infective, lethally hot, a liquid that would scare the daylights out of a military biohazard specialist. The smell of the *vomito negro* fills the passenger cabin. The airsickness bag is brimming with black vomit, so Monet closes the bag and rolls up the top. The bag is bulging and softening, threatening to leak, and he hands it to a flight attendant.

When a hot virus multiplies in a host, it can saturate the body with virus particles, from the brain to the skin. The military experts then say that the virus has undergone ''extreme amplification.'' This is not something like the common cold. By the time an extreme amplification peaks out, an eyedropper of the victim's blood may contain a hundred million particles of virus. During this process, the body is partly transformed into virus particles. In other words, the host is possessed by a life form that is attempting to convert the host into *itself.* The transformation is not entirely successful, however, and the end result is a great deal of liquefying flesh mixed with virus, a kind of biological accident. Extreme amplification has occurred in Monet, and the sign of it is the black vomit.

He appears to be holding himself rigid, as if any movement would rupture something inside him. His blood is clotting up—his bloodstream is throwing clots, and the clots are lodging everywhere. His liver, kidneys, lungs, hands, feet, and head are becoming jammed with blood clots. In effect, he is having a stroke through the whole body. Clots are accumulating in his intestinal muscles, cutting off the blood supply to his intestines. The intestinal muscles are beginning to die, and the intestines are starting to go slack. He doesn't seem to be fully aware of pain any longer because the blood clots lodged in his brain are cutting off blood flow. His personality is being wiped away by brain damage. This is called depersonalization, in which the liveliness and details of character seem to vanish. He is becoming an automaton. Tiny spots in his brain are liquefying. The higher functions of consciousness are winking out first, leaving the deeper parts of the brain stem (the primitive rat brain, the lizard brain) still alive and functioning. It could be said that the *who* of Charles Monet has already died while the *what* of Charles Monet continues to live.

The vomiting attack appears to have broken some blood vessels in his nose—he gets a nosebleed. The blood comes from both nostrils, a shining, clotless, arterial liquid that drips over his teeth and chin. This blood keeps running, because the clotting factors have been used up. A flight attendant gives him some paper towels, which he uses to stop up his nose, but the blood still won't coagulate, and the towels soak through.

When a man is getting sick in an airline seat next to you, you may not want to embarrass him by calling attention to the problem. You say to yourself that this man will be all right. Maybe he doesn't travel well in airplanes. He is airsick, the poor man, and people do get nosebleeds in airplanes, the air is so dry and thin . . . and you ask him, weakly, if there is anything you can do to help. He does not answer, or he mumbles words you can't understand, so you try to ignore it, but the flight seems to go on forever. Perhaps the flight attendants offer to help him. But victims of this type of hot virus have changes in behavior that can render them incapable of responding to an offer of help. They become hostile, and don't want to be touched. They don't want to speak. They answer questions with grunts or monosyllables. They can't seem to find words. They can tell you their name, but they can't tell you the day of the week or explain what has happened to them.

The Friendship drones through the clouds, following the length of the Rift Valley, and Monet slumps back in the seat, and now he seems to be dozing. . . . Perhaps some of the passengers wonder if he is dead. No, no, he is not dead. He is moving. His red eyes are open and moving around a little bit.

It is late afternoon, and the sun is falling down into the hills to the west of the Rift Valley, throwing blades of light in all directions, as if the sun is cracking up on the equator. The Friendship makes a gentle turn and crosses the eastern scarp of the Rift. The land rises higher and changes in color from brown to green. The Ngong Hills appear under the right wing, and the plane, now descending, passes over parkland dotted with zebra and giraffes. A minute later, it lands at Jomo Kenyatta International Airport. Monet stirs himself. He is still able to walk. He stands up, dripping. He stumbles down the gangway onto the tarmac. His shirt is a red mess. He carries no luggage. His only luggage is internal, and it is a load of amplified virus. Monet has been transformed into a human virus bomb. He walks slowly into the airport terminal and through the building and out to a curving road where taxis are always parked. The taxi drivers surround him— "Taxi?" "Taxi?"

"Nairobi . . . Hospital," he mumbles.

One of them helps him into a car. Nairobi taxi drivers like to chat with their fares, and this one probably asks if he is sick. The answer should be obvious. Monet's stomach feels a little better now. It is heavy, dull, and bloated, as if he has eaten a meal, rather than empty and torn and on fire.

The taxi pulls out onto the Uhuru Highway and heads into Nairobi. It goes through grassland studded with honey-acacia trees, and it goes past factories, and then it comes to a rotary and enters the bustling street life of Nairobi. Crowds are milling on the shoulders of the road, women walking on beaten dirt pathways, men loitering, children riding bicycles, a man repairing shoes by the side of the road, a tractor pulling a wagonload of charcoal. The taxi turns left onto the Ngong Road and goes past a city park and up a hill, past lines of tall blue-gum trees, and it turns up a narrow road and goes past a guard gate and enters the grounds of Nairobi Hospital. It parks at a taxi stand beside a flower kiosk. A sign by a glass door says CASUALTY DEPT. Monet hands the driver some money and gets out of the taxi and opens the glass door and goes over to the reception window and indicates that he is very ill. He has difficulty speaking.

The man is bleeding, and they will admit him in just a moment. He must wait until a doctor can be called, but the doctor will see him immediately, not to worry. He sits down in the waiting room.

It is a small room lined with padded benches. The clear, strong, ancient light of East Africa pours through a row of windows and falls across a table heaped with soiled magazines, and makes rectangles on a pebbled gray floor that has a drain in the center. The room smells vaguely of wood-smoke and sweat, and it is jammed with bleary-eyed people, Africans and Europeans sitting shoulder to shoulder. There is always someone in Casualty who has a cut and is waiting for stitches. People wait patiently, holding a washcloth against the scalp, holding a bandage pressed around a finger, and you may see a spot of blood on the cloth. So Charles Monet is sitting on a bench in Casualty, and he does not look very much different from anyone else in the room, except for his bruised, expressionless face and his red eyes. A sign on the wall warns patients to watch out for purse thieves, and another sign says:

PLEASE MAINTAIN **SILENCE.**
YOUR COOPERATION WILL BE APPRECIATED.
NOTE: THIS IS A CASUALTY DEPARTMENT.
EMERGENCY CASES WILL BE TAKEN IN PRIORITY.
YOU MAY BE REQUIRED TO WAIT FOR SUCH CASES
BEFORE RECEIVING ATTENTION.

Monet maintains silence, waiting to receive attention. Suddenly he goes into the last phase—the human virus bomb explodes. Military biohazard specialists have ways of describing this occurrence. They say that the victim has "crashed and bled out." Or more politely they say that the victim has "gone down."

He becomes dizzy and utterly weak, and his spine goes limp and nerveless and he loses all sense of balance. The room is turning around and around. He is going into shock. He leans over, head on his knees, and brings up an incredible quantity of blood from his stomach and spills it onto the floor with a gasping groan. He loses consciousness and pitches forward onto the floor. The only sound is a choking in his throat as he continues to vomit blood and black matter while unconscious. Then comes a sound like a bedsheet being torn in half, which is the sound of his bowels opening and venting blood from the anus. The blood is mixed with intestinal lining. He has sloughed his gut. The linings of his intestines have come off and are being expelled along with huge amounts of blood. Monet has crashed and is bleeding out.

The other patients in the waiting room stand up and move away from the man on the floor, calling for a doctor. Pools of blood spread out around him, enlarging rapidly. Having destroyed its host, the hot agent is now coming out of every orifice, and is "trying" to find a new host.

Jumper

1980 JANUARY 15

Nurses and aides came running, pushing a gurney along with them, and they lifted Charles Monet onto the gurney and wheeled him into the intensive care unit at Nairobi Hospital. A call for a doctor went out over the loudspeakers: a patient was bleeding in the ICU. A young doctor named Shem Musoke ran to the scene. Dr. Musoke was widely considered to be one of the best young physicians at the hospital, an energetic man with a warm sense of humor, who worked long hours and had a good feel for emergencies. He found Monet lying on the gurney. He had no idea what was wrong with the man, except that he was obviously having some kind of massive hemorrhage. There was no time to try to figure out what had caused it. He was having difficulty breathing—and then his breathing stopped. He had inhaled blood and had had a breathing arrest.

Dr. Musoke felt for a pulse. It was weak and sluggish. A nurse ran and fetched a laryngoscope, a tube that can be used to open a person's airway. Dr. Musoke ripped open Monet's shirt so that he could observe any rise and fall of the chest, and he stood at the head of the gurney and bent over Monet's face until he was looking directly into his eyes, upside down.

Monet stared redly at Dr. Musoke, but there was no movement in the eyeballs, and the pupils were dilated. Brain damage: nobody home. His nose was bloody and his mouth was bloody. Dr. Musoke tilted the patient's head back to open the airway so that he could insert the laryngoscope. He was not wearing rubber gloves. He ran his finger around the patient's tongue to clear the mouth of debris, sweeping out mucus and blood. His hands became greasy with black curd. The patient smelled of vomit and blood, but this was nothing new to Dr. Musoke, and he concentrated on his work. He leaned down until his face was a few inches away from Monet's face, and he looked into Monet's mouth in order to judge the position of the scope. Then he slid the scope over Monet's tongue and pushed the tongue out of the way so that he could see down the airway past the epiglottis, a dark hole leading inward to the lungs. He pushed the scope into the hole, peering into the instrument. Monet suddenly jerked and thrashed.

Monet vomited.

The black vomit blew up around the scope and out of Monet's mouth. Black-and-red fluid spewed into the air, showering down over Dr. Musoke. It struck him in the eyes. It splattered over his white coat and down his chest, marking him with strings of red slime dappled with dark flecks. It landed in his mouth.

He repositioned his patient's head and swept the blood out of the patient's mouth with his fingers. The blood had covered Dr. Musoke's hands, wrists, and forearms. It had gone everywhere—all over the gurney, all over Dr. Musoke, all over the floor. The nurses in the intensive care unit couldn't believe their eyes. Dr. Musoke peered down into the airway and pushed the scope deeper into the lungs. He saw that the airways were bloody.

Air rasped into the man's lungs. The patient had begun to breathe again.

The patient was apparently in shock from loss of blood. He had lost so much blood that he was becoming dehydrated. The blood had come out of practically every opening in his body. There wasn't enough blood left to maintain circulation, so his heartbeat was very sluggish, and his blood pressure was dropping toward zero. He needed a blood transfusion.

A nurse brought a bag of whole blood. Dr. Musoke hooked the bag on a stand and inserted the needle into the patient's arm. There was something wrong with the patient's veins; his blood poured out around the needle. Dr. Musoke tried again, putting the needle into another place in the patient's arm and probing for the vein. Failure. More blood poured out. At every place in the patient's arm where he stuck the needle, the vein broke apart like cooked macaroni and spilled blood, and the blood ran from the punctures down the patient's arm and wouldn't coagulate. Clearly his blood was not normal. Dr. Musoke abandoned his efforts to give his patient a blood transfusion for fear that the patient would bleed to death out of the small hole in his arm. The patient continued to bleed from the bowels, and these hemorrhages were now as black as pitch.

Monet's coma deepened, and he never regained consciousness. He died in the intensive care unit in the early hours of the morning. Dr. Musoke stayed by his bedside the whole time.

They had no idea what had killed him. It was an unexplained death. They opened him up for an autopsy and found that his kidneys were destroyed and that his liver was destroyed. It was yellow, and parts of it had liquefied—it looked like the liver of a cadaver. It was as if Monet had become a corpse before his death. Sloughing of the gut, in which the intestinal lining comes off, is another effect that is ordinarily seen in a corpse that is several days old. What, exactly, was the cause of death? It was impossible to say because there were too many possible causes. Everything had gone wrong inside this man, absolutely everything, any one of which could have been fatal: the clotting, the massive hemorrhages, the liver turned into pudding, the intestines full of blood. Lacking words, categories, or language to describe what had happened, they called it, finally, a case of "fulminating liver failure." His remains were placed in a waterproof bag and, according to one account, were buried locally. When I visited Nairobi, years later, no one remembered where the grave was.

1980 JANUARY 24

Nine days after the patient vomited into Dr. Shem Musoke's eyes and mouth, Musoke developed an aching sensation in his back. He was not prone to backaches—really, he had never had a serious backache—but he was approaching thirty, and it occurred to him that he was getting into the time of life when some men begin to get bad backs. He had been driving himself hard these past few weeks. He had been up all night with a patient who had had heart problems, and then, the following night, he had been up most of the night with that Frenchman with hemorrhages who had come from somewhere upcountry. So he had been going nonstop for days without sleep. He hadn't thought much about the vomiting incident, and when the ache began to spread through his body, he still didn't think about it. Then, when he looked in a mirror, he noticed that his eyes were turning red.

Red eyes—he began to wonder if he had malaria. He had a fever now, so certainly he had some kind of infection. The backache had spread until all the muscles in his body ached badly. He started taking malaria pills, but they didn't do any good, so he asked one of the nurses to give him an injection of an antimalarial drug.

The nurse gave it to him in the muscle of his arm. The pain of the injection was very, very bad. He had never felt such pain from a shot; it was abnormal and memorable. He wondered why a simple shot would give him this kind of pain. Then he developed abdominal pain, and that made him think that he might have typhoid fever, so he gave himself a course of antibiotic pills, but that had no effect on his illness. Meanwhile, his patients needed him, and he continued to work at the hospital. The pain in his stomach and in his muscles grew unbearable, and he developed jaundice.

Unable to diagnose himself, in severe pain, and unable to continue with his work, he presented himself to Dr. Antonia Bagshawe, a physician at Nairobi Hospital. She examined him, observed his fever, his red eyes, his jaundice, his abdominal pain, and came up with nothing definite, but wondered if he had gallstones or a liver abscess. A gall-bladder attack or a liver abscess could cause fever and jaundice and abdominal pain—the red eyes she could not explain—and she ordered an ultrasound examination of his liver. She studied the images of his liver and saw that it was enlarged, but, other than that, she could see nothing unusual. By this time, he was very sick, and they put him in a private room with nurses attending him around the clock. His face set itself into an expressionless mask.

This possible gallstone attack could be fatal. Dr. Bagshawe recommended that Dr. Musoke have exploratory surgery. He was opened up in the main operating theater at Nairobi Hospital by a team of surgeons headed by Dr. Imre Lofler. They made an incision over his liver and pulled back the abdominal muscles. What they found inside Musoke was eerie and disturbing, and they could not explain it. His liver was swollen and red and did not look healthy, but they could not find any sign of gallstones. Meanwhile, he would not stop bleeding. Any surgical procedure will cut through blood vessels, and the cut vessels will ooze for a while and then clot up, or if the oozing continues, the surgeon will put dabs of gel foam on them to stop the bleeding. Musoke's blood vessels would not stop oozing—his blood would not clot. It was as if he had become a hemophiliac. They dabbed gel foam all over his liver, and the blood came through the foam.

He leaked blood like a sponge. They had to suction off a lot of blood from the incision, but as they pumped it out, the incision filled up again. It was like digging a hole below the water table: it fills up as fast as you pump it out. One of the surgeons would later tell people that the team had been "up to the elbows in blood." They cut a wedge out of his liver—a liver biopsy—and dropped the wedge into a bottle of pickling fluid and closed up Musoke as quickly as they could.

He deteriorated rapidly after the surgery, and his kidneys began to fail. He appeared to be dying. At that time, Antonia Bagshawe, his physician, had to travel abroad, and he came under the care of a doctor named David Silverstein. The prospect of kidney failure and dialysis for Dr. Musoke created a climate of emergency at the hospital—he was well liked by his colleagues, and they didn't want to lose him. Silverstein began to suspect that Musoke was suffering from an unusual virus. He collected some blood from his patient and drew off the serum, which is a clear, golden-colored liquid that remains when the red cells are removed from the blood. He sent some tubes of frozen serum to laboratories for testing—to the National Institute of Virology in Sandringham, South Africa, and to the Centers for Disease Control in Atlanta, Georgia, U.S.A. Then he waited for results.

Diagnosis

David Silverstein lives in Nairobi, but he owns a house near Washington, D.C. One day in the summer recently, when he was visiting the United States to tend to some business, I met him in a coffee shop in a shopping mall not far from his home. We sat at a small table, and he told me about the Monet and Musoke cases. Silverstein is a slender, short man in his late forties, with a mustache and glasses, and he has an alert, quick gaze. Although he is an American, his voice carries a hint of a Swahili accent. On the day that I met him, he was dressed in a denim jacket and blue jeans, and he was nicely tanned, looking fit and relaxed. He is a pilot, and he flies his own plane. He has the largest private medical practice in East Africa, and it has made him a famous figure in Nairobi. He is the personal physician of Daniel arap Moi, the president of Kenya, and he travels with President Moi when Moi goes abroad. He treats all the important people in East Africa: the corrupt politicians, the actors and actresses who get sick on safari, the decayed English-African nobility. He traveled at the side of Diana, Lady Delamere, as her personal physician when she was growing old, to monitor her blood pressure and heartbeat (she wanted to carry on with her beloved sport of deep-sea fishing off the Kenya coast, although she had a heart condition), and he was also Beryl Markham's doctor. Markham, the author of *West with the*

Night, a memoir of her years as an aviator in East Africa, used to hang out at the Nairobi Aero Club, where she had a reputation for being a slam-bang, two-fisted drinker. ("She was a well-pickled old lady by the time I came to know her.") His patient Dr. Musoke has himself become a celebrity, in the annals of disease. "I was treating Dr. Musoke with supportive care," Silverstein said to me. "That was all I could do. I tried to give him nutrition, and I tried to lower his fevers when they were high. I was basically taking care of somebody without a game plan."

One night, at two o'clock in the morning, Silverstein's telephone rang at his home in Nairobi. It was an American researcher stationed in Kenya calling him to report that the South Africans had found something very queer in Musoke's blood: "He's positive for Marburg virus. This is really serious. We don't know much about Marburg."

Silverstein had never heard of Marburg virus. "After the phone call, I could not get back to sleep," he said to me. "I had kind of a waking dream about it, wondering what Marburg was." He lay in bed, thinking about the sufferings of his friend and colleague Dr. Musoke, fearful of what sort of organism had gotten loose among the medical staff at the hospital. He kept hearing the voice saying, "We don't know much about Marburg." Unable to sleep, he finally got dressed and drove to the hospital, arriving at his office before dawn. He found a medical textbook and looked up Marburg virus.

The entry was brief. Marburg is an African organism, but it has a German name. Viruses are named for the place where they are first discovered. Marburg is an old city in central Germany, surrounded by forests and meadows, where factories nestle in green valleys. The virus erupted there in 1967, in a factory called the Behring Works, which produced vaccines using kidney cells from African green monkeys. The Behring Works regularly imported monkeys from Uganda. The virus came to Germany hidden somewhere in a series of air shipments of monkeys totaling five or six hundred animals. As few as two or three of the animals were incubating the virus. They were probably not even visibly sick. At any rate, shortly after they arrived at the Behring Works, the virus began to spread among them, and a few of them crashed and bled out. Soon afterward, the Marburg agent jumped species and suddenly emerged in the human population of the city. This is an example of virus amplification.

The first person known to be infected with the Marburg agent was a man called Klaus F., an employee at the Behring Works vaccine factory who fed the monkeys and washed their cages. He broke with the virus on August 8, 1967, and died two weeks later. So little is known about the Marburg agent that only one book has been published about it, a collection of papers presented at a symposium on the virus, held at the University of Marburg in 1970. In the book, we learn that

> The monkey-keeper HEINRICH P. came back from his holiday on August 13th 1967 and did his job of killing monkeys from the 14th–23rd. The first symptoms appeared on August 21st.
>
> The laboratory assistant RENATE L. broke a test-tube that was to be sterilized, which had contained infected material, on August 28th, and fell ill on September 4th 1967.

And so on. The victims developed headaches at about day seven after their exposure and went downhill from there, with raging fevers, clotting, spurts of blood, and terminal shock. For a few days in Marburg, doctors in the city thought the world was coming to an end. Thirty-one people eventually caught the virus; seven died in pools of blood. The kill rate of Marburg turned out to be about one in four, which makes Marburg an extremely lethal agent: even in the best modern hospitals, where the patients are hooked up to life-support machines, Marburg kills a quarter of the patients who are infected with it. By contrast, yellow fever, which is considered a highly lethal virus, kills only about one in twenty patients once they reach a hospital.

Marburg is one of a family of viruses known as the filoviruses. Marburg was the first filovirus to be discovered. The word *filovirus* is Latin and means "thread virus." The filoviruses look alike, as if they are sisters, and they resemble no other virus on earth. While most viruses are ball-shaped particles that look like peppercorns, the thread viruses have been compared to strands of tangled rope, to hair, to worms, to snakes. When they appear in a great flooding mess, as they so often do when they have destroyed a victim, they look like a tub of spaghetti that has been dumped on the floor. Marburg particles sometimes roll up into loops. The loops resemble Cheerios. Marburg is the only ring-shaped virus known.

In Germany, the effects of Marburg virus on the brain were particularly frightening, and resembled the effects of rabies: the virus somehow damaged the central nervous system and could destroy the brain, as does rabies. The Marburg particles also looked rather like rabies particles. The rabies-virus particle is shaped like a bullet. If you stretch out a bullet, it begins to look like a length of rope, and if you coil the rope into a loop, it becomes a ring, like Marburg. Thinking that Marburg might be related to rabies, they called it stretched rabies. Later it became clear that Marburg belongs to its own family.

Not long after Charles Monet died, it was established that the family of filoviruses comprised Marburg along with two types of a virus called Ebola. The Ebolas were named Ebola Zaire and Ebola Sudan. Marburg was the mildest of the three filovirus sisters. The worst of them was Ebola Zaire. The kill rate in humans infected with Ebola Zaire is nine out of ten. Ninety percent of the people who come down with Ebola Zaire die of it. Ebola Zaire is a slate wiper in humans.

Marburg virus (the gentle sister) affects humans some-what like nuclear radiation, damaging virtually all of the tissues in their bodies. It attacks with particular ferocity the internal organs, connective tissue, intestines, and skin. In Germany, all the survivors lost their hair—they went bald or partly bald. Their hair died at the roots and fell out in clumps, as if they had received radiation burns. Hemorrhage occurred from all orifices of the body. I have seen a photograph of one of the men who died of Marburg, taken in the hours before his death. He is lying in bed without any clothing on his upper body. His face is expressionless. His chest, arms, and face are speckled with blotches and bruises, and droplets of blood stand on his nipples.

During the survivors' recovery period, the skin peeled off their faces, hands, feet, and genitals. Some of the men suffered from blown up, inflamed, semirotten testicles. One of the worst cases of this testicular infection appeared in a morgue attendant who had handled Marburg-infected bodies and who himself came down with Marburg, having caught it from the cadavers. The virus also lingered in the fluid inside the eyeballs of some victims for many months. No one knows why Marburg has a special affinity for the testicles and the eyes. One man infected his wife with Marburg through sexual intercourse.

Doctors noticed that the Marburg agent had a strange effect on the brain. "Most of the patients showed a sullen, slightly aggressive, or negativistic behavior," according to the book. "Two patients [had] a feeling as if they were lying on crumbs." One patient became psychotic, apparently as a result of brain damage. The patient called Hans O.-V. showed no signs of mental derangement, and his fever cooled, and he seemed to be stabilizing, but then suddenly, without warning, he had an acute fall in blood pressure—he was crashing—and he died. They performed an autopsy on him, and when they opened his skull, they found a massive, fatal hemorrhage at the center of the brain. He had "bled out" into his brain.

International health authorities were urgently concerned

to find the exact source of the monkeys, in order to pin down
where in nature the Marburg virus lived. It seemed pretty
clear that the Marburg virus did not naturally circulate in
monkeys, because it killed them so fast it could not successfully establish itself in them as a useful host. Therefore,
Marburg lived in some other kind of host—an insect? a rat?
a spider? a reptile? Where, exactly, had the monkeys been
trapped? That place would be the hiding place of the virus.
Soon after the outbreak in Germany, a team of investigators
under the auspices of the World Health Organization flew to
Uganda to try to find out where those monkeys had come
from. It turned out they had been trapped at locations all
over central Uganda. The team couldn't discover the exact
source of the virus.

There the mystery lingered for many years. Then, in 1982,
an English veterinarian came forward with new eyewitness
information about the Marburg monkeys. I will call this man
Mr. Jones (today, he prefers to remain anonymous). During
the summer of 1967, when the virus erupted in Germany,
Mr. Jones was working at a temporary job inspecting monkeys at the export facility in Entebbe from which the sick
Marburg monkeys had been shipped, while the regular veterinary inspector was on leave. This monkey house, which was
run by a rich monkey trader ("a sort of lovable rogue,"
according to Mr. Jones) was exporting about thirteen thousand monkeys a year to Europe. This was a very large number of monkeys, and it generated big money. The infected
shipment was loaded onto an overnight flight to London, and
from there it was flown to Germany—where the virus broke
out of the monkeys and "attempted" to establish itself in the
human population.

After making a number of telephone calls, I finally located Mr. Jones in a town in England, where today he is
working as a veterinary consultant. He said to me: "All that
the animals got, before they were shipped off, was a visual
inspection."

"By whom?" I asked.

"By me," he said. "I inspected them to see that they appeared normal. On occasion, with some of these shipments, one or two animals were injured or had skin lesions." His method was to pick out the sick-looking ones, which were removed from the shipment and presumably killed before the remaining healthy-looking animals were loaded onto the plane. When, a few weeks later, the monkeys started the outbreak in Germany, Mr. Jones felt terrible. "I was appalled, because I had signed the export certificate," he said to me. "I feel now that I have the deaths of these people on my hands. But that feeling suggests I could have done something about it. There was no way I could have known." He is right about that: the virus was then unknown to science, and as few as two or three not-visibly-sick animals could have started the outbreak. One concludes that the man should not be blamed for anything.

The story becomes more disturbing. He went on: "The sick ones were being killed, or so I thought." But later he learned that they weren't being killed. The boss of the company was having the sick monkeys put in boxes and shipped out to a small island in Lake Victoria, where they were being released. With so many sick monkeys running around it, the island could have become a focus for monkey viruses. It could have been a hot island, an isle of plagues. "Then, if this guy was a bit short of monkeys, he went out to the island and caught a few, unknown to me, and those infected or recently infected monkeys were then shipped off to Europe." Mr. Jones thinks it is possible that the Marburg agent had established itself on the hot island, and was circulating among the monkeys there, and that some of the monkeys which ended up in Germany had actually come from that island. But when the WHO team came later to investigate, "I was told by my boss to say nothing unless asked." As it turned out, no one asked Mr. Jones any questions—he says he never met the WHO team. The fact that the team apparently never spoke with him, the monkey inspector, "was bad epidemiology but good politics," he remarked to me. If it had been revealed that the monkey trader was shipping off suspect monkeys collected on a suspect island, he could have been put out of business, and Uganda would have lost a source of valuable foreign cash.

Shortly after the Marburg outbreak in Germany, Mr. Jones recalled a fact that began to seem important to him. It seems that the Marburg virus may have been burning through rural areas in Uganda not far from Kitum Cave. Between 1962 and 1965 he had been stationed in eastern Uganda, on the slopes of Mount Elgon, inspecting cattle for diseases. At some time during that period, local chiefs told him that the people who lived on the north side of the volcano, along the Greek River, were suffering from a disease that caused bleeding, death, and "a peculiar skin rash"— and that monkeys in the area were dying of a similar disease. Mr. Jones did not pursue the rumors, and was never able to confirm the nature of the disease. But it seems possible that in the years preceding the outbreak of Marburg virus in Germany, a hidden outbreak of the virus occurred on the slopes of Mount Elgon.

Mr. Jones's personal vision of the Marburg outbreak reminds me of a flashlight pointed down a dark hole. It gives a narrow but disturbing view of the larger phenomenon of the origin and spread of tropical viruses. He told me that some of the Marburg monkeys were trapped in a group of islands in Lake Victoria known as the Sese Islands. The Seses are a low-lying forested archipelago in the northwestern part of Lake Victoria, an easy boat ride from Entebbe. The isle of plagues may have been situated among the Seses or near them. Mr. Jones does not recall the name of the hot island. He says it is "close" to Entebbe. At any rate, Mr. Jones's then-boss, the Entebbe monkey trader, had arranged a deal with villagers in the Sese Islands to buy monkeys from them. They regarded the monkeys as pests and were happy to get rid of them, especially for money. So the trader was obtaining wild monkeys from the Sese Islands, and if the animals proved to be sick, he was releasing them again on another island somewhere near Entebbe. And some monkeys from the isle of plagues seemed to be ending up in Europe.

In papyrus reeds and desolate flatlands on the western shore of Lake Victoria facing the Sese Islands, there is a fishing village called Kasensero. You can see the Sese Islands from the village. Kasensero was one of the first places in the world where AIDS appeared. Epidemiologists have since discovered that the northwestern shore of Lake Victoria was one of the initial epicenters of AIDS. It is generally believed that AIDS came originally from African primates, from monkeys and apes, and that it somehow jumped out of these animals into the human race. It is thought that the virus went through a series of very rapid mutations at the time of its jump from primates to humans, which enabled it to establish itself successfully in people. In the years since the AIDS virus emerged, the village of Kasensero has been devastated. The virus has killed a large portion of the inhabitants. It is said that other villages along the shores of Lake Victoria have been essentially wiped off the map by AIDS.

The villagers of Kasensero are fishermen who were, and are, famous as smugglers. In their wooden boats and motorized canoes they ferried illegal goods back and forth across the lake, using the Sese Islands as hiding places. One can guess that if a monkey trader were moving monkeys around Lake Victoria, he might call on the Kasensero smugglers or on their neighbors.

One general theory for the origin of AIDS goes that, during the late nineteen-sixties, a new and lucrative business grew up in Africa, the export of primates to industrialized countries for use in medical research. Uganda was one of the

biggest sources of these animals. As the monkey trade was
established throughout central Africa, the native workers in
the system, the monkey trappers and handlers, were exposed
to large numbers of wild monkeys, some of which were
carrying unusual viruses. These animals, in turn, were being
jammed together in cages, exposed to one another, passing
viruses back and forth. Furthermore, different species of
monkeys were mixed together. It was a perfect setup for an
outbreak of a virus that could jump species. It was also a
natural laboratory for rapid virus evolution, and possibly it
led to the creation of HIV. Did HIV crash into the human race
as a result of the monkey trade? Did AIDS come from an
island in Lake Victoria? A hot island? Who knows. When
you begin probing into the origins of AIDS and Marburg, the
light fails and things go dark, but you sense hidden connec-
tions. Both viruses seem part of a pattern.

When he learned what Marburg virus does to human beings,
Dr. David Silverstein persuaded the Kenyan health authorities to shut down Nairobi Hospital. For a week, patients who
arrived at the doors were turned away, while sixty-seven
people were quarantined inside the hospital, mostly medical
staff. They included the doctor who had done the autopsy on
Monet, nurses who had attended Monet or Dr. Musoke, the
surgeons who had operated on Musoke, and aides and technicians who had handled any secretions from either Monet
or Musoke. It turned out that a large part of the hospital's
staff had had direct contact with either Monet or Musoke or
with blood samples and fluids that came from the two patients. The surgeons who had operated on Musoke, remembering only too well that they had been "up to the elbows in
blood," sweated in quarantine for two weeks while they
wondered if they were going to break with Marburg. A single human virus bomb had walked into the hospital's waiting
room and exploded there, and the event had put the hospital

out of business. Charles Monet had been an Exocet missile that struck the hospital below the water line.

Dr. Shem Musoke survived his encounter with a hot agent. Ten days after he fell sick, the doctors noticed a change for the better. Instead of merely lying in bed in a passive state, he became disoriented and angry and refused to take medicine. One day, a nurse was trying to turn him over in bed, and he waved his fist at her and cried, "I have a stick, and I will beat you." It was around that time that he began to get better, and after many days his fever subsided and his eyes cleared; his mind and personality came back, and he recovered slowly but completely. Today he is one of the leading physicians at Nairobi Hospital, where he practices as a member of David Silverstein's group. One day I interviewed him, and he said to me that he has almost no memory of the weeks he was infected with Marburg. "I only remember bits and pieces," he said. "I remember having major confusion. I remember, before my surgery, that I walked out of my room with my IV drip hanging out of me. I remember the nurses just turning me and turning me in bed. I don't remember much of the pain. The only pain I can talk about is the muscle ache and the lower-back ache. And I remember him throwing up on me." Nobody else at the hospital developed a proven case of Marburg-virus disease.

When a virus is trying, so to speak, to crash into the human species, the warning sign may be a spattering of breaks at different times and places. These are microbreaks. What had happened at Nairobi Hospital was an isolated emergence, a microbreak of a rain-forest virus with unknown potential to start an explosive chain of lethal transmission in the human race.

Tubes of Dr. Musoke's blood went to laboratories around the world so that they could have samples of living Marburg for their collections of life forms. The Marburg in his blood had come from Charles Monet's black vomit and perhaps originally from Kitum Cave. Today this particular strain of Marburg virus is known as the Musoke strain. Some of it ended up in glass vials in freezers owned by the United States Army, where it was kept immortal in a zoo of hot agents.

A Woman
and a Soldier

Thurmont, Maryland, nearly four years after the death of
Charles Monet. Evening. A typical American town. On
Catoctin Mountain, a ridge of the Appalachians that runs
north to south through the western part of the state, the trees
were brightening into soft yellows and golds. Teenagers
drove their pickup trucks slowly along the streets of the
town, looking for something to happen, wishing that the
summer had not ended. Faint smells of autumn touched
the air, the scent of ripening apples, a sourness of dead
leaves, cornstalks drying in the fields. In the apple groves at
the edge of town, flocks of grackles settled into the branches
for the night, squawking. Headlights streamed north on the
Gettysburg road.

In the kitchen of a Victorian house near the center of
town, Major Nancy Jaax, a veterinarian in the United States
Army, stood at a counter making dinner for her children. She
slid a plate into the microwave oven and pushed a button.
Time to nuke up some chicken for the kids. Nancy Jaax wore
sweatpants and a T-shirt, and she was barefoot. Her feet had
calluses on them, the result of martial-arts training. She had
wavy auburn hair, which was cut above the shoulders, and
greenish eyes. Her eyes were actually two colors, green with

an inner rim around the iris that was amber. She was a former homecoming queen from Kansas—Miss Agriculture, Kansas State. She had a slender, athletic build, and she displayed quick motions, flickery gestures, with her arms and hands. Her children were restless and tired, and she worked as fast as she could to fix the dinner.

Jaime, who was five, hung on Nancy's leg. She grabbed the leg of Nancy's sweatpants and pulled, and Nancy lurched sideways, and then Jaime pulled the other way, and Nancy lurched to the other side. Jaime was short for her age and had greenish eyes, like her mother. Nancy's son, Jason, who was seven, was watching television in the living room. He was rail thin and quiet, and when he grew up he would probably be tall, like his father.

Nancy's husband, Major Gerald Jaax, whom everyone called Jerry, was also a veterinarian. He was in Texas at a training class, and Nancy was alone with the children. Jerry had telephoned to say that it was hot as hell in Texas, and he missed her badly and wished he was home. She missed him, too. They had not been apart for more than a few days at a time ever since they had first started dating, in college.

Nancy and Jerry Jaax—the name is pronounced JACKS— were both members of the Army Veterinary Corps, a tiny corps of "doggy doctors." They take care of the Army's guard dogs, as well as Army horses, Army cows, Army sheep, Army pigs, Army mules, Army rabbits, Army mice, and Army monkeys. They also inspect the Army's food.

Nancy and Jerry had bought the Victorian house not long after they had been assigned to Fort Detrick, which was nearby, within easy commuting distance. The kitchen was very small, and at the moment you could see plumbing and wires hanging out of the walls. Not far from the kitchen, the living room had a bay window with a collection of tropical plants and ferns in it, and there was a cage among the plants that held an Amazon parrot named Herky. The parrot burst into a song:

Heigh-ho, heigh-ho,
it's home from work we go!

"Mom! Mom!" he cried excitedly. His voice sounded like Jason's.

"What?" Nancy said. Then she realized it was the parrot. "Nerd brain," she muttered.

The parrot wanted to sit on Nancy's shoulder. "Mom! Mom! Jerry! Jaime! Jason!" the parrot shouted, calling everyone in the family. When he didn't get any response, he whistled the "Colonel Bogey March" from *The Bridge on the River Kwai*. And then: "Whaat? Whaat? Mom! *Mom!*"

Nancy did not want to take Herky out of his cage. She worked quickly, putting plates and silverware out on the counter. Some of the officers at Fort Detrick had noticed a certain abrupt quality in her hand motions and had accused her of having hands that were "too quick" to handle delicate work in dangerous situations. Nancy had begun martial-arts training partly because she hoped to make her gestures cool and smooth and powerful, and also because she had felt the frustrations of a woman officer trying to advance her career in the Army. She was five feet four inches tall. She liked to spar with six-foot male soldiers, big guys. She enjoyed knocking them around a little bit; it gave her a certain satisfaction to be able to kick higher than the guy's head. She used her feet more than her hands when she sparred with an opponent, because her hands were delicate. She could break four boards with a spinning back kick. She had reached the point where she could kill a man with her bare feet, an idea that did not in itself give her much satisfaction. On occasion, she had come home from her class with a broken toe, a bloody nose, or a black eye. Jerry would just shake his head: Nancy with another shiner.

Major Nancy Jaax did all the housework. She could not stand housework. Scrubbing grape jelly out of rugs didn't give her a feeling of reward, and in any case she did not have time for it. Occasionally she would go into a paroxysm of cleaning, and she would race around the house for an hour, throwing things into closets. She also did all the cooking for her family. Jerry was useless in the kitchen. Another point of contention was his tendency to buy things impulsively—a motorcycle, a sailboat. Jerry had bought a sailboat when they were stationed at Fort Riley in Kansas. And then there was that god-awful diesel Cadillac with a red leather interior. She and Jerry had commuted to work together in it, but the car had started to lay smoke all over the road even before the payments were finished. One day, she had finally said to Jerry, ''You can sit in the driveway in those red leather seats all you want, but I'm not getting in there with you.'' So they sold the Cadillac and bought a Honda Accord.

The Jaaxes' house was the largest Victorian house in town, a pile of turreted brick with a slate roof and tall windows and a cupola and wooden paneling made of golden American chestnut. It stood on a street corner near the ambulance station. The sirens woke them up at night. They had bought the house cheap. It had sat on the market a long time, and a story had been going around town that the previous owner had hanged himself in the basement. After the Jaaxes bought it, the dead man's widow showed up at the door one day. She was a wizened old lady, come to have a look around her old place, and she fixed a blue eye on Nancy and said, ''Little girl, you are going to hate this house. I did.''

There were other animals in the house besides the parrot. In a wire cage in the living room lived a python named Sampson. He would occasionally escape from his cage, wander around the house, and eventually climb up inside the hollow center post of the dining-room table and go to sleep. There he would stay for a few days. It gave Nancy a creepy feeling to think that there was a python asleep inside the dining table. You wondered whether the snake was going to wake up while you were eating dinner. Nancy had a study in the cupola at the top of the house. The snake had once escaped from his cage and disappeared for a few days. They pounded and knocked on the dining-room table to try to flush him out, but he wasn't there. Late one night when Nancy was in her study, the snake oozed out of the rafters and hung in front of her face, staring at her with lidless eyes, and she screamed. The family also had an Irish setter and an Airedale terrier. Whenever the Jaaxes were assigned to a different Army post, the animals moved with them in boxes and cages, a portable ecosystem of the Jaax family.

Nancy loved Jerry. He was tall and fine looking, a handsome man with prematurely gray hair. She thought of his hair as silver, to go along with his silver tongue, which he used trying to talk her into buying diesel Cadillacs with red-leather interiors. He had sharp brown eyes and a sharp nose, like a hawk's, and he understood her better than anyone else on earth. Nancy and Jerry Jaax had very little social life outside of their marriage. They had grown up on farms in Kansas, twenty miles apart as the crow flies, but had not known each other as children. They met in veterinary school at Kansas State University and had gotten engaged a few weeks later, and they were married when Nancy was twenty. By the time they graduated, they were broke and in debt, with no money to set up a practice as veterinarians, and so they had enlisted in the Army together.

Since Nancy didn't have time to cook during the week, she would spend her Saturdays cooking. She would make up a beef stew in a Crock-Pot, or she would broil several chickens. Then she would freeze the food in bags. On weekday nights, she would take a bag out of the freezer and heat it in the microwave. Tonight, while she thawed chicken, she considered the question of vegetables. How about canned green beans? The children liked that. Nancy opened a cabinet and pulled down a can of Libby's green beans.

She searched through one or two drawers, looking for a

can opener. Couldn't find it. She turned to the main junk drawer, which held all the utensils, the stirring spoons and vegetable peelers. It was a jam-packed nightmare.

The hell with the can opener. She pulled a butcher knife out of the drawer. Her father had always warned her not to use a knife to open a can, but Nancy Jaax had never made a point of listening to her father's advice. She jabbed the butcher knife into the can, and the point stuck in the metal. She hit the handle with the heel of her right hand. All of a sudden her hand slipped down the handle, struck the tang of the blade, and slid down the blade. She felt the edge bite deep.

The butcher knife clattered to the floor, and big drops of blood fell on the counter. "Son of a bitch!" she said. The knife had sliced through the middle of her right hand, on the palm. It was a deep cut. She wondered if the knife had hit bone or cut any tendons. She put pressure on the cut to stanch the bleeding and went over to the sink, turned on the faucet, and thrust her hand under the stream of water. The sink turned red. She wiggled her fingers. They worked; so she had not sliced a tendon. This was not such a bad cut. Holding her hand over her head, she went into the bathroom and found a Band-Aid. She waited for the blood to coagulate, and then she pressed the Band-Aid over the cut, drawing the sides of the cut together to seal the wound. She hated the sight of blood, even if it was her own blood. She had a thing about blood. She knew what some blood could contain.

Nancy skipped the children's baths because of the cut on her hand and gave them their usual snuggle in bed. That night, Jaime slept in bed with her. Nancy didn't mind, especially because Jerry was out of town, and it made her feel close to her children. Jaime seemed to need the reassurance. Jaime was always a little edgy when Jerry was out of town.

Project Ebola

The next morning, Nancy Jaax woke up at four o'clock. She got out of bed quietly so as not to wake Jaime and showered and put on her uniform. She wore green Army slacks with a black stripe down the leg, a green Army shirt, and in the cold before sunrise she put on a black military sweater. The sweater displayed the shoulder bars of a major, with gold oak leaves. She drank a Diet Coke to wake herself up, and walked upstairs to her study in the cupola of the house.

Today she might put on a biohazard space suit. She was in training for veterinary pathology, the study of disease in animals. Her speciality was turning out to be the effects of Biosafety Level 4 hot agents, and in the presence of those kinds of agents you need to wear a space suit. She was also studying for her pathology-board exams, which were coming up in a week. As the sun rose that morning over the apple orchards and fields to the east of town, she opened her books and hunched over them. Grackles began croaking in the trees, and trucks began to move along the streets of Thurmont, below her window. The palm of her right hand still throbbed.

At seven o'clock, she went down to the master bedroom

and woke Jaime, who was curled up in the bed. She went into Jason's room. Jason was harder to wake, and Nancy had to shake him several times. Then the babysitter arrived, an older woman named Mrs. Trapane, who got Jaime and Jason dressed and gave them their breakfasts while Nancy climbed back up to the cupola and returned to her books. Mrs. Trapane would see Jason off to the school bus and would watch Jaime at home until Nancy came back from work that evening.

At seven-thirty, Nancy closed her books and kissed her children good-bye. She thought to herself, Have to remember to stop at the bank and get some money to pay Mrs. Trapane. She drove the Honda alone to work, heading south on the Gettysburg road, along the foot of Catoctin Mountain. As she approached Fort Detrick, in the city of Frederick, the traffic thickened and slowed. She turned off the highway and arrived at the main gate of the base. A guard waved her through. She turned right, drove past the parade grounds with its flagpole, and parked her car near a massive, almost windowless building made of concrete and yellow bricks that covered almost ten acres of ground. Tall vent pipes on the roof discharged filtered exhaust air that was being pumped out of sealed biological laboratories inside the building. This was the United States Army Medical Research Institute of Infectious Diseases, or USAMRIID.

Military people often call USAMRIID the Institute. When they call the place USAMRIID, they drawl the word in a military way, making it sound like "you Sam rid," which gives it some hang time in the air. The mission of USAMRIID is medical defense. The Institute conducts research into ways to protect soldiers against biological weapons and natural infectious diseases. It specializes in drugs, vaccines, and biocontainment. At the Institute, there are always a number of programs going on simultaneously—research into vaccines for various kinds of bacteria, such as anthrax and botulism, research into the characteristics of viruses that might infect American troops, either naturally or in the form of a battlefield weapon. Beginning with the Second World War, Army labs at Fort Detrick performed research into offensive biological weapons—the Army was developing strains of lethal bacteria and viruses that could be loaded into bombs and dropped on an enemy. In 1969, President Richard M. Nixon signed an executive order that outlawed the development of offensive biological weapons in the United States. From then on, the Army labs were converted to peaceful uses, and USAMRIID was founded. It devoted itself to developing protective vaccines, and it concentrated on basic research into ways to control lethal microorganisms. The Institute knows ways to stop a monster virus before it ignites an explosive chain of lethal transmission in the human race.

Major Nancy Jaax entered the building through the back entrance and showed her security badge to a guard behind a desk, who nodded and smiled at her. She headed into the main block of containment zones, traveling through a maze of corridors. There were soldiers everywhere, dressed in fatigues, and there were civilian scientists and technicians wearing ID badges. People seemed very busy, and rarely did anyone stop to chat with someone else in the corridors.

Nancy wanted to see what had happened to the Ebola monkeys during the night. She walked along a Biosafety Level 0 corridor, heading for a Level 4 biocontainment area known as AA-5, or the Ebola suite. The levels are numbered 0, 2, 3 and finally 4, the highest. (For some reason, there is no Level 1.) All the containment levels at the Institute, from Level 2 to Level 4, are kept under negative air pressure, so that if a leak develops, air will flow *into* the zones rather than outward to the normal world. The suite known as AA-5 was a group of negative-pressure biocontainment rooms that had been set up as a research lab for Ebola virus by a civilian Army scientist named Eugene Johnson. He was an expert in Ebola and its sister, Marburg. He had infected several monkeys with Ebola virus, and he had been giving them various drugs to see if they would stop the Ebola infection. In recent days, the monkeys had begun to die. Nancy had joined Johnson's Ebola project as the pathologist. It was her job to determine the cause of death in the monkeys.

She came to a window in a wall. The window was made of heavy glass, like that in an aquarium, and it looked directly into the Ebola suite, directly into Level 4. You could not see the monkeys through this window. Every morning, a civilian animal caretaker put on a space suit and went inside to feed the monkeys and clean their cages and check on their physical condition. This morning there was a piece of paper taped to the inside of the glass, with handwritten lettering on it. It had been left there by a caretaker. The note said that during the night two of the animals had "gone down." That is, they had crashed and bled out.

When she saw the note, she knew that she would be putting on a space suit and going in to dissect the monkeys. Ebola virus destroyed an animal's internal organs, and the carcass deteriorated abruptly after death. It softened, and the tissues turned into jelly, even if you put it in a refrigerator to keep it cold. You wanted to dissect the animals quickly, before the spontaneous liquefaction began, because you can't dissect gumbo.

When Nancy Jaax first applied to join the pathology group at the Institute, the colonel in charge of it didn't want to accept her. Nancy thought it was because she was a woman. He said to her, "This work is not for a married female. You are either going to neglect your work or neglect your family." One day, she brought her résumé into his office, hoping to persuade him to accept her. He said, "I can have anybody I want in my group"—implying that he didn't want her because she wasn't good enough—and he mentioned the great Thoroughbred stallion Secretariat. "If I want to have Secretariat in my group," he said, "I can get Secretariat."

"Well, sir, I am no plow horse!" she roared at him, and

slammed her résumé on his desk. He reconsidered the matter and allowed her to join the group.

When you begin working with biological agents, the Army starts you in Biosafety Level 2, and then you move up to Level 3. You don't go into Level 4 until you have a lot of experience, and the Army may never allow you to work there. In order to work in the lower levels, you must have a number of vaccinations. Nancy had vaccinations for yellow fever, Q fever, Rift Valley fever, the VEE, EEE, and WEE complex (brain viruses that live in horses), and tularemia, anthrax, and botulism. And, of course, she had had a series of shots for rabies, since she was a veterinarian. Her immune system reacted badly to all the shots: they made her sick. The Army therefore yanked her out of the vaccination program. At this point, Nancy Jaax was essentially washed up. She couldn't proceed with any kind of work with Level 3 agents, because she couldn't tolerate the vaccinations. There was only one way she could continue working with dangerous infectious agents. She had to get herself assigned to work in a space suit in Level 4 areas. There aren't any vaccines for Level 4 hot agents. A Level 4 hot agent is a lethal virus for which there is no vaccine and no cure.

Ebola virus is named for the Ebola River, which is the headstream of the Mongala River, a tributary of the Congo, or Zaire, River. The Ebola River empties tracts of rain forest, winding past scattered villages. The first known emergence of Ebola Zaire—the hottest type of Ebola virus—occurred in September 1976, when it erupted simultaneously in fifty-five villages near the headwaters of the Ebola River. It seemed to come out of nowhere, and killed nine out of ten people it infected. Ebola Zaire is the most feared agent at the Institute. The general feeling around USAMRIID has always been "Those people who work with Ebola are crazy." To mess around with Ebola is an easy way to die. Better to work with something safer, such as anthrax.

Eugene Johnson, the civilian biohazard expert who was running the Ebola research program at the Institute, had a reputation for being a little bit wild. He is something of a legend to the handful of people in the world who really know about hot agents and how to handle them. He is one of the world's leading Ebola hunters. Gene Johnson is a large man, not to say massive, with a broad, heavy face and loose-flying disheveled brown hair and a bushy brown beard and a gut that hangs over his belt, and glaring, deep eyes. If Gene Johnson were to put on a black leather jacket, he could pass for a roadie with the Grateful Dead. He does not look at all like a man who works for the Army. He has a reputation for being a top-notch field epidemiologist (a person who studies viral diseases in the wild), but for some reason he does not often get around to publishing his work. That explains his somewhat mysterious reputation. When people who know Johnson's work talk about him, you hear things like "Gene Johnson did this, Gene Johnson did that," and it all sounds clever and imaginative. He is a rather shy man, somewhat suspicious of people, deeply suspicious of viruses. I think I have never met someone who is more afraid of viruses than Gene Johnson, and what makes his fear impressive is the fact that it is a deep intellectual respect, rooted in knowledge. He spent years traveling across central Africa in search of the reservoirs of Ebola and Marburg viruses. He had virtually ransacked Africa looking for these life forms, but despite his searches he had never found them in their natural hiding places. No one knew where any of the filoviruses came from; no one knew where they lived in nature. The trail had petered out in the forests and savannas of central Africa. To find the hidden reservoir of Ebola was one of Johnson's great ambitions.

No one around the Institute wanted to get involved with his Ebola project. Ebola, the slate wiper, did things to people that you did not want to think about. The organism was too frightening to handle, even for those who were comfortable and adept in space suits. They did not care to do research on Ebola because they did not want Ebola to do research on them. They didn't know what kind of host the virus lived in —whether it was a fly or a bat or a tick or a spider or a scorpion or some kind of reptile, or an amphibian, such as a frog or a newt. Or maybe it lived in leopards or elephants. And they didn't know exactly how the virus spread, how it jumped from host to host.

Gene Johnson had suffered recurrent nightmares about Ebola virus ever since he began to work with it. He would wake up in a cold sweat. His dreams went more or less the same way. He would be wearing his space suit while holding Ebola in his gloved hand, holding some kind of liquid tainted with Ebola. Suddenly the liquid would be running all over his glove, and then he would realize that his glove was full of pinholes, and the liquid was dribbling over his bare hand and running inside his space suit. He would come awake with a start, saying to himself, *My God, there's been an exposure.* And then he would find himself in his bedroom, with his wife sleeping beside him.

In reality, Ebola had not yet made a decisive, irreversible breakthrough into the human race, but it seemed close to doing that. It had been emerging in microbreaks here and there in Africa. The worry was that a microbreak would develop into an unstoppable tidal wave. If the virus killed nine out of ten people it infected and there was no vaccine or cure for it, you could see the possibilities. The possibilities were global. Johnson liked to say to people that we don't really know what Ebola has done in the past, and we don't know what it might do in the future. Ebola was unpredictable. An airborne strain of Ebola could emerge and circle around the world in about six weeks, like the flu, killing large numbers of people, or it might forever remain a secret feeder at the margins, taking down humans a few at a time.

Ebola is a rather simple virus—as simple as a firestorm. It kills humans with swift efficiency and with a devastating range of effects. Ebola is distantly related to measles, mumps, and rabies. It is also related to certain pneumonia viruses: to the parainfluenza virus, which causes colds in children, and to the respiratory syncytial virus, which can cause fatal pneumonia in a person who has AIDS. In its own evolution through unknown hosts and hidden pathways in the rain forest, Ebola seems to have developed the worst elements of all the above viruses. Like measles, it triggers a rash all over the body. Some of its effects resemble rabies—psychosis, madness. Other of its effects look eerily like a bad cold.

The Ebola virus particle contains only seven different proteins—seven distinct types of large molecules arranged in a long braided structure that is the stringy Ebola particle. Three of these proteins are vaguely understood, and four of the proteins are completely unknown—their structure and their function is a mystery. Whatever these Ebola proteins do, they seem to target the immune system for special attack. In this they are like HIV, which also destroys the immune system, but unlike the creeping onset of HIV, the attack by Ebola is explosive. As Ebola sweeps through you, your immune system fails, and you seem to lose your ability to respond to viral attack. Your body becomes a city under seige, with its gates thrown open and hostile armies pouring in, making camp in the public squares and setting everything on fire; and from the moment Ebola enters your bloodstream, the war is already lost; you are almost certainly doomed. You can't fight off Ebola the way you fight off a cold. Ebola does in ten days what it takes AIDS ten years to accomplish.

It is not really known how Ebola is transmitted from person to person. Army researchers believed that Ebola virus traveled through direct contact with blood and bodily fluids (in the same way the AIDS virus travels). Ebola seemed to have other routes of travel as well. Many of the people in Africa who came down with Ebola had handled Ebola-infected cadavers. It seems that one of Ebola's paths goes from the dead to the living, winding in trickles of uncoagulated blood and slimes that come out of the dead body. In Zaire during the 1976 outbreak, grieving relatives kissed and embraced the dead or prepared the body for burial, and then, three to fourteen days later, they broke with Ebola.

Gene Johnson's Ebola experiment was simple. He would infect a few monkeys with the virus, and then he would treat them with drugs in the hope that they would get better. That way, he might discover a drug that would fight Ebola virus or possibly cure it.

Monkeys are nearly identical to human beings in a biological sense, which is why they are used in medical experiments. Humans and monkeys are both primates, and Ebola feeds on primates in the same way that a predator consumes certain kinds of flesh. Ebola can't tell the difference between a human being and a monkey. The virus jumps easily back and forth between them.

Nancy Jaax volunteered to work as the pathologist on Johnson's Ebola project. It was Level 4 work, which she was qualified to do, because she didn't need to be vaccinated. She was eager to prove herself and eager to continue working with lethal viruses. However, some people around the Institute were skeptical of her ability to work in a space suit in Level 4. She was a "married female"—and therefore, they claimed, she might panic. They claimed that her hands looked nervous or clumsy, not good for work with Level 4 hot agents. People felt that she might cut herself or stick herself with a contaminated needle—or stick someone else. Her hands became a safety issue. But the real issue was that she was a woman.

Her immediate superior was Lieutenant Colonel Anthony Johnson (he is not related to Gene Johnson, the civilian who was the head of the Ebola project). Tony Johnson is a soft-spoken man and a cool customer. Now he had to decide

whether to allow her to go into Biosafety Level 4. Wanting to
be sure he understood the situation, he sent word around the
Institute: Who knows Nancy Jaax? Who can comment on
her strengths and weaknesses? Major Jerry Jaax, Nancy's
husband, showed up in Lieutenant Colonel Johnson's office.
Jerry was against the idea of his wife putting on a space suit.
He argued strongly against it. He said that there had been
"family discussions" about Nancy working with Ebola virus. "Family discussions." Jerry had said to Nancy, "You're
the only wife I've got" . . . He did not wear a biological
space suit himself at work, and he did not want his wife to
wear one either. His biggest concern was that she would be
handling Ebola. He could not stand the idea that his wife, the
woman he loved, the mother of their children, would hold in
her hands a monstrous life form that is lethal and incurable.

Lieutenant Colonel Tony Johnson listened to what Major
Jerry Jaax had to say, and listened to what other people had
to say, and then he felt he should speak with Nancy himself,
and so he called her into his office. He could see that she was
tense. He watched her hands as she talked. They looked fine
to him, not clumsy, and not too quick, either. He decided
that the rumors he had been hearing about her hands were
unfounded. She said to him, "I don't want any special favors." Well, she was not going to get any special favors.
"I'm going to put you in the Ebola program," he said. He
told her that he would allow her to put on a space suit and go
into the Ebola area, and that he would accompany her on the
first few trips, to teach her how to do it and to observe her
hands at work. He would watch her like a hawk. He believed
that she was ready for total immersion in a hot zone.

As he spoke, she broke down and cried in front of him—
"had a few tears," as he would later recall. They were tears
of happiness. At that moment, to hold Ebola virus in her
hands was what she wanted more than anything else in the
world.

1300 HOURS

Nancy spent the morning doing paperwork in her office. After lunch, she removed her diamond engagement ring and her wedding band and locked them in her desk drawer. She dropped by Tony Johnson's office and asked him if he was ready to go in. They went downstairs and along a corridor to the Ebola suite. There was only one locker room leading into it. Tony Johnson insisted that Nancy Jaax go in first, to get changed. He would follow.

The room was small and contained a few lockers along one wall, some shelves, and a mirror over a sink. She undressed, removing all of her clothing, including her underwear, and put everything in her locker. She left the Band-Aid stuck to her hand. From a shelf, she took up a sterile surgical scrub suit—green pants and a green shirt, the clothing that a surgeon wears in an operating room—and she dragged on the pants and tied the drawstring at the waist, and snapped the shirt's snaps. You were not allowed to wear anything under the scrub suit, no underwear. She pulled a cloth surgical cap over her head and tucked her hair up into the cap while looking in the mirror. She did not appear nervous, but she was starting to feel a little bit nervous. This was only her second trip into a hot area.

Standing in her bare feet, she turned away from the mirror and faced a door leading into Level 2. A deep blue light streamed through a window in the door—ultraviolet light. Viruses fall apart under ultraviolet light, which smashes their genetic material and makes them unable to replicate.

As she opened the door and entered Level 2, she felt the door stick against her pull, sucked in by a difference of air pressure, and a gentle drag of air whispered around her shoulders and traveled inward, toward the hot zone. This was negative air pressure, designed to keep hot agents from drifting outward. The door closed behind her, and she was in

Level 2. The blue light bathed her face. She walked through a water-shower stall that contained an ultraviolet light, a bar of soap, and some ordinary shampoo. The shower stall led into a bathroom, where there was a shelf that held some clean white socks. She put on a pair of socks and pushed through another door, into Level 3.

This was a room known as the staging area. It contained a desk with a telephone and a sink. A cylindrical waxed cardboard box sat on the floor beside the desk. It was a biohazard container known as a "hatbox," or "ice-cream container." A hatbox is blazed with biohazard symbols, which are red, spiky three-petaled flowers, and it is used for storing and transporting infectious waste. This hatbox was empty. It was only a makeshift chair.

She found a box of latex rubber surgical gloves and a plastic shaker full of baby powder. She shook baby powder onto her hands and pulled on the gloves. Then she found a roll of sticky tape, and she tore off several strips of tape and hung them in a row on the edge of the desk. Then she taped herself. Taking up one strip at a time, she taped the cuffs of her gloves to the sleeves of her scrub shirt, running the tape around the cuff to make a seal. She then taped her socks to her trousers. Now she wore one layer of protection between herself and the replicative Other.

Lieutenant Colonel Johnson came in through Level 2 wearing a surgical scrub suit. He put on rubber gloves and began taping them to his sleeves, and he taped his socks to his pants.

Nancy turned to the right, into an antechamber, and found her space suit hanging on a rack. It was a Chemturion biological space suit, and it was marked in letters across the chest: JAAX. The Chemturion type is also known as a blue suit, because it is bright blue. It is a pressurized, heavy-duty plastic space suit that meets government specifications for work with airborne hot agents.

She opened up the space suit and laid it down on the

concrete floor and stepped into it, feet first. She pulled it up
to her armpits and slid her arms into the sleeves until her
fingers entered the gloves. The suit had brown rubber gloves
that were attached by gaskets at the cuffs. These were the
space suit's main gloves, and they were made of heavy rubber. They were the most important barrier between her and
Ebola. The hands were the weak point, the most vulnerable
part of the suit, because of what they handled. They handled
needles, knives, and sharp pieces of bone. You are responsible for maintaining your space suit in the same way that a
paratrooper is responsible for packing and maintaining his
own parachute. Perhaps Nancy was in a bit of a hurry and
did not inspect her space suit as closely as she should have.

Lieutenant Colonel Johnson gave her a short briefing on
procedures and then helped her lower the helmet over her
head. The helmet was made of soft, flexible plastic. Johnson
looked at her face, visible through the clear faceplate, to see
how she was doing.

She closed an oiled Ziploc zipper across the suit's chest.
The zipper made a popping sound as it snapped shut, *pop,
pop, pop*. The moment the space suit was closed, her
faceplate fogged up. She reached over to a wall and pulled
down a coiled yellow air hose and plugged it into her suit.
Then came a roar of flowing air, and her suit bloated up, fat
and hard, and a whiff of dry air cleared away some tiny
beads of sweat that had collected inside her faceplate.

Around the Institute, they say that you can't predict who
will panic inside a biological space suit. It happens now and
then, mainly to inexperienced people. The moment the helmet closes over their faces, their eyes begin to glitter with
fear, they sweat, turn purple, claw at the suit, try to tear it
open to get some fresh air, lose their balance and fall down
on the floor, and they can start screaming or moaning inside
the suit, which makes them sound as if they are suffocating
in a closet. There was one case in which a man in Level 4
suddenly began screaming, "Get me out of here!"—and he

tore off his space suit's helmet, taking great gasps of air from Level 4. (They dragged him into a chemical shower and kept him there for a while.)

After he had helped Nancy Jaax put on her space suit, and had looked into her eyes for signs of panic, Tony Johnson put on his own suit, and when he was closed up and ready, he handed her a pack of dissection tools. He seemed calm and collected. They turned and faced the stainless-steel door together. The door led into an air lock and Level 4. The door was plastered with a biohazard symbol and warnings:

CAUTION

BIOHAZARD

DO NOT ENTER
WITHOUT WEARING VENTILATED SUIT

The international symbol for biohazard, which is pasted on doors at USAMRIID whenever they open through a major transition of zones, is a red trefoil that reminds me of a red trillium, or toadshade.

The Level 4 air lock is a gray area, a place where two worlds meet, where the hot zone touches the normal world. The gray area is neither hot nor cold. A place that is neither provably sterile nor known to be infective. At USAMRIID, toadshades bloom in the gray zones. Nancy took a breath and gathered her thoughts into stillness, using her martial-arts training to get her breathing under control. People performed all kinds of small rituals before they walked through that steel door. Some people crossed themselves. Others carried amulets or charms inside their space suits, even though it was technically against the rules to bring anything inside the suit except your body and the surgical scrubs. They hoped the amulets might help ward off the hot agent if there was a major break in their suit.

She unplugged her air hose and unlatched the steel door and entered an air lock, and Tony Johnson followed her. The air lock was made entirely of stainless steel, and it was lined with nozzles for spraying water and chemicals. This was the decon shower. *Decon* means "decontamination." The door closed behind them. Nancy opened the far door of the air lock, and they crossed over to the hot side.

Total Immersion

They were standing in a narrow cinder-block corridor. Various rooms opened on either side. The hot zone was a maze. From the walls dangled yellow air hoses. There was an alarm strobe light on the ceiling that would be triggered if the air system failed. The walls were painted with thick, gobby epoxy paint, and all the electrical outlets were plugged around the edges with a gooey material. This was to seal any cracks and holes, so that a hot agent could not escape by drifting through hollow electrical conduits. Nancy reached for an air hose and plugged it into her suit. She could not hear anything except the roar of air in her helmet. The air rumbled so loudly in their suits that they did not try to speak to each other.

She opened a metal cabinet. Blue light streamed out of it, and she removed a pair of yellow rubber boots. They reminded her of barn boots. She slid the soft feet of her space suit into the boots and glanced at Johnson and caught his eye. Ready for action, boss.

They unplugged their air hoses and proceeded down the hallway and entered the monkey room. It contained two banks of cages, positioned facing each other along opposite walls of the room. Jaax and Johnson replugged their hoses and peered into the cages. One bank of cages contained two isolated monkeys. They were the so-called control monkeys. They had not been injected with Ebola virus, and they were healthy.

As soon as the two Army officers appeared in space suits, the healthy monkeys went nuts. They rattled their cages and leaped around. Humans in space suits make monkeys nervous. They hooted and grunted—*"Ooo! Ooo! Haw, wah, haw!"* And they uttered a high-pitched squeal: *"Eek!"* The monkeys moved to the front of their cages and shook the doors or leaped back and forth, *whump, whump, whump,* watching Jaax and Johnson the whole time, following them with their eyes, alert to everything. The cages had elaborate bolts on the doors to prevent fiddling by primate fingers. These monkeys were creative little boogers, she thought, and they were bored.

The other bank of cages was mostly quiet. This was the bank of Ebola cages. All the monkeys in these cages were infected with Ebola virus, and most of them were silent, passive, and withdrawn, although one or two of them seemed queerly deranged. Their immune systems had failed or gone haywire. Most of the animals did not look very sick yet, but they did not display the alertness, the usual monkey energy, the leaping and the cage rattling that you see in healthy monkeys, and most of them had not eaten their morning biscuits. They sat almost motionless in their cages, watching the two officers with expressionless faces.

They had been injected with the hottest strain of Ebola known to the world. It was the Mayinga strain of Ebola Zaire. This strain had come from a young woman named Mayinga N., who died of the virus on October 19, 1976. She was a nurse at a hospital in Zaire, and she had taken care of a Roman Catholic nun who died of Ebola. The nun had bled to death all over Nurse Mayinga, and then, a few days later, Nurse Mayinga had broken with Ebola and died. Some of Nurse Mayinga's blood had ended up in the United States,

and the strain of virus that had once lived in Nurse Mayinga's blood now lived in small glass vials kept in super-
freezers at the Institute, which were maintained at minus one
hundred and sixty degrees Fahrenheit. The freezers were fitted with padlocks and alarms and were plastered with biohazard flowers and sealed with bands of sticky tape. The first
line of defense against a hot agent is sticky tape, because it
seals cracks. It could be said that without sticky tape there
would be no such thing as biocontainment.

Gene Johnson, the civilian scientist, had thawed a little bit
of Nurse Mayinga's frozen blood and had injected it into the
monkeys. Then, as the monkeys became sick, he had treated
them with a drug in the hope that it would help them fight off
the virus. The drug did not seem to be working.

Nancy Jaax and Tony Johnson inspected the monkeys,
moving from cage to cage, until they found the two monkeys
that had crashed and bled out. Those animals were hunched
up, each in its own cage. They had bloody noses, and their
eyes were half-open, glassy, and brilliant red, with dilated
pupils. The monkeys showed no facial expression, not even
pain or agony. The connective tissue under the skin had been
destroyed by the virus, causing a subtle distortion of the
face. Another reason for the strange faces was that the parts
of the brain that control facial expression had also been
destroyed. The masklike face, the red eyes, and the bloody
nose were classic signs of Ebola that appear in all primates
infected with the virus, both monkeys and humans. It hinted
at a vicious combination of brain damage and soft-tissue
destruction under the skin. The classic Ebola face made the
monkeys look as if they had seen something beyond compre-
hension. It was not a vision of heaven.

Nancy Jaax felt a wave of unease. She was distressed by
the sight of the dead and suffering monkeys. As a veterinarian, she believed that it was her duty to heal animals and
relieve their suffering. As a scientist, she believed that it was
her obligation to perform medical research that would help

alleviate human suffering. Even though she had grown up on a farm, where her father had raised livestock for food, she had never been able to bear easily the death of an animal. As a girl, she had cried when her father had taken her 4-H Club prize steer to the butcher. She liked animals better than many people. In taking the veterinarian's oath, she had pledged herself to a code of honor that bound her to the care of animals but also bound her to the saving of human lives through medicine. At times in her work, those two ideals clashed. She told herself that this research was being done to help find a cure for Ebola, that it was medical research that would help save human lives and might possibly avert a tragedy for the human species. That helped reduce her feelings of unease, but not completely, and she kept her emotions off to one side.

Johnson watched Jaax carefully as she began the removal procedure. Handling an unconscious monkey in Level 4 is a tricky operation, because monkeys can wake up, and they have teeth and a powerful bite, and they are remarkably strong and agile. The monkeys that are used in laboratories are not organ-grinder monkeys. They are large, wild animals from the rain forest. A bite by an Ebola-infected monkey would almost certainly be fatal.

First Nancy inspected the monkey, looking through the bars. It was a large male, and he looked as if he was really dead. She saw that he still had his canine fangs, and that made her nervous. Ordinarily the monkey would have had its fangs filed down for safety. For some reason, this one had enormous natural fangs. She stuck her gloved fingers through the bars and pinched the monkey's toe while she watched for any eye movement. The eyes remained fixed and staring.

"GO AHEAD AND UNLOCK THE CAGE," Lieutenant Colonel Johnson said. He had to shout to be heard above the roar of air in their space suits.

She unlocked the door and slid it up until the cage gaped

wide open. She inspected the monkey again. No muscle twitches. The monkey was definitely down.

"ALL RIGHT, GO AHEAD AND MOVE HIM OUT," Johnson said.

She reached inside and caught the monkey by the upper arms and rotated him so that he was facing away from her, so that he couldn't bite her if he woke up. She pulled his arms back and held them immobile, and she lifted the monkey out of the cage.

Johnson took the monkey's feet, and together they carried him over to a hatbox, a biohazard container, and they slid the monkey into it. Then they carried the hatbox to the necropsy room, shuffling slowly in their suits. They were two human primates carrying another primate. One was the master of the earth, or at least believed himself to be, and the other was a nimble dweller in trees, a cousin of the master of the earth. Both species, the human and the monkey, were in the presence of another life form, which was older and more powerful than either of them, and was a dweller in blood.

Jaax and Johnson moved slowly out of the room, carrying the monkey, and turned left and then turned left again, and entered the necropsy room, and laid the monkey down on a stainless-steel table. The monkey's skin was rashy and covered with red blotches, visible through his sparse hair.

"GLOVE UP," Johnson said.

They put on latex rubber gloves, pulling them over the space-suit gloves. They now wore three layers of gloves: the inner-lining glove, the space-suit glove, and the outer glove. Johnson said, "WE'LL DO THE CHECK LIST. SCISSORS. HEMOSTATS." He laid the tools in a row at the head of the table. Each tool was numbered, and he called the numbers out loud.

They went to work. Using blunt-ended scissors, Johnson opened the monkey while Jaax assisted with the procedure. They worked slowly and with exquisite care. They did not use any sharp blades, because a blade is a deadly object in a hot zone. A scalpel can nick your gloves and cut your fingers, and before you even feel a sensation of pain, the agent has already entered your bloodstream.

Nancy handed tools to him, and she reached her fingers inside the monkey to tie off blood vessels and mop up excess blood with small sponges. The animal's body cavity was a lake of blood. It was Ebola blood, and it had run everywhere inside the animal: there had been a lot of internal hemorrhaging. The liver was swollen, and she noticed some blood in the intestines.

She had to tell herself to slow her hands down. Perhaps her hands were moving too quickly. She talked herself through the procedure, keeping herself alert and centered. Keep it clean, keep it clean, she thought. Okay, pick up the hemostat. Clamp that artery 'cause it's leaking blood. Break off and rinse gloves. She could feel the Ebola blood through her gloves: it felt wet and slippery, even though her hands were clean and dry and dusted with baby powder.

She withdrew her hands from the carcass and rinsed them in a pan of disinfectant called EnviroChem, which sat in a sink. The liquid was pale green, the color of Japanese green tea. It destroys viruses. As she rinsed her gloves in it, the liquid turned brown with monkey blood. All she could hear was the noise of the air blowing inside her suit. It filled her suit with a roar like a subway train coming through a tunnel.

A virus is a small capsule made of membranes and proteins. The capsule contains one or more strands of DNA or RNA, which are long molecules that contain the software program for making a copy of the virus. Some biologists classify viruses as "life forms," because they are not strictly known to be alive. Viruses are ambiguously alive, neither alive nor dead. They carry on their existence in the borderlands between life and nonlife. Viruses that are outside cells merely sit there; nothing happens. They are dead. They can even form crystals. Virus particles that lie around in blood or mucus may seem dead, but the particles are waiting for something to come along. They have a sticky surface. If a cell comes along and touches the virus and the stickiness of the virus matches the stickiness of the cell, then the virus clings to the cell. The cell feels the virus sticking to it and enfolds the virus and drags it inside. Once the virus enters the cell, it becomes a Trojan horse. It switches on and begins to replicate.

A virus is a parasite. It can't live on its own. It can only make copies of itself inside a cell using the cell's materials and machinery to get the job done. All living things carry viruses in their cells. Even fungi and bacteria are inhabited by viruses and are occasionally destroyed by them. That is, diseases have their own diseases. A virus makes copies of itself inside a cell until eventually the cell gets pigged with virus and pops, and the viruses spill out of the broken cell. Or viruses can bud through a cell wall, like drips coming out of a faucet—drip, drip, drip, drip, copy, copy, copy, copy— that's the way the AIDS virus works. The faucet runs and runs until the cell is exhausted, consumed, and destroyed. If enough cells are destroyed, the host dies. A virus does not "want" to kill its host. That is not in the best interest of the virus, because then the virus may also die, unless it can jump fast enough out of the dying host into a new host.

The genetic code inside Ebola is a single strand of RNA. This type of molecule is thought to be the oldest and most "primitive" coding mechanism for life. The earth's primordial ocean, which came into existence not long after the earth was formed, about four and a half billion years ago, may well have contained microscopic life forms based on RNA. This suggests that Ebola is an ancient kind of life, perhaps nearly as old as the earth itself. Another hint that Ebola is extremely ancient is the way in which it can seem neither quite alive nor quite unalive.

Viruses may seem alive when they multiply, but in another sense they are obviously dead, are only machines, subtle ones to be sure, but strictly mechanical, no more alive than a jackhammer. Viruses are molecular sharks, a motive without a mind. Compact, hard, logical, totally selfish, the virus is dedicated to making copies of itself—which it can do on occasion with radiant speed. The prime directive is to replicate.

Viruses are too small to be seen. Here is a way to imagine the size of a virus. Consider the island of Manhattan shrunk to this size:

\

This Manhattan could easily hold nine million viruses. If you could magnify this Manhattan and if it were full of viruses, you would see little figures clustered like the lunch crowd on Fifth Avenue. A hundred million crystallized polio viruses could cover the period at the end of this sentence. There could be two hundred and fifty Woodstock Festivals of viruses sitting on that period—the combined populations of Great Britain and France—and you would never know it.

Keep it clean, Nancy thought. No blood. No blood. I *don't like* blood. Every time I see a drop of blood, I see a billion viruses. Break off and rinse. Break off and rinse. Slow yourself. Look at Tony's suit. Check him.

You watched your partner's suit for any sign of a hole or a break. It was kind of like being a mother and watching your kid—a constant background check to see if everything is okay.

Meanwhile, Johnson was checking her. He observed her for any kind of mistake, any jerkiness with the tools. He wondered if he would see her drop something.

"RONGEUR," he said.

"WHAT?" she asked.

He pointed at her air hose to suggest that she crimp it so

that she could hear him better. She grabbed the hose and folded it. The air stopped flowing, her suit deflated around her, and the noise died away. He put his helmet close to hers and spoke the word *rongeur* again, and she released her hose. She handed him a pair of pliers called the rongeur. The word is French and means "gnawer." It is used for opening skulls.

Getting into a skull is always a bitch in Level 4. A primate skull is hard and tough, and the bone plates are knitted together. Ordinarily you would whip through a skull with an electric bone saw, but you can't use a bone saw in Level 4. It would throw a mist of bone particles and blood droplets into the air, and you do not want to create any kind of infective mist in a hot area, even if you are wearing a space suit; it is just too dangerous.

They popped the skull with the pliers. It made a loud cracking sound. They removed the brain, eyes, and spinal cord and dropped them into a jar of preservative.

Johnson was handing her a tube containing a sample when he stopped and looked at her gloved hands. He pointed to her right glove.

She glanced down. Her glove. It was drenched in blood, but now she saw the hole. It was a rip across the palm of the outer glove on her right hand.

Nancy tore off the glove. Now her main suit glove was covered with blood. It spidered down the outer sleeve of her space suit. Great, just great—Ebola blood all over my suit. She rinsed her glove and arm in the disinfectant, and they came up clean and shiny wet. Then she noticed that her hand, inside the two remaining gloves, felt cold and clammy. There was something wet inside her space-suit glove. She wondered if that glove was a leaker, too. She wondered if she had sustained a breach in her right main glove. She inspected that glove carefully. Then she saw it. It was a crack in the wrist. She had a breach in her space suit. Her hand felt *wet*.

She wondered if there might be Ebola blood inside her space suit, somewhere close to that cut on the palm of her hand. She pointed to her glove and said, "HOLE." Johnson bent over and inspected her glove. He saw the crack in the wrist. She saw his face erupt in surprise, and then he looked into her eyes. She saw that he was afraid.

That terrified her. She jerked her thumb toward the exit. "I'M OUTTA HERE, MAN. CAN YOU FINISH?"

He replied, "I WANT YOU TO LEAVE IMMEDI-ATELY. I'LL SECURE THE AREA AND FOLLOW YOU OUT."

Using only her left hand, her good hand, she unplugged her suit from the air hose. She practically ran down the corridor to the air lock, her right arm hanging rigidly at her side. She did not want to move that hand because every time she moved it she felt something squishing around in there, inside the glove. Fear threatened to overwhelm her. How was she going to remove her boots without using her bad hand? She kicked them off. They went flying down the corridor. She threw open the air-lock door and stepped inside and slammed the door behind her.

She pulled a chain that hung from the ceiling of the air lock. That started the decon shower. The decon shower takes seven minutes, and you are not permitted to leave during that period, because the shower needs time to work on viruses. First came a blast of water jets, which washed traces of blood from her space suit. The water jets stopped. Then came a spray of EnviroChem, coming out of nozzles all up and down the sides of the air lock, which deconned her space suit. Of course, if something lived inside her glove, the chemical shower would not reach it.

There were no lights in the air lock; it was dim, almost dark. The place was literally a gray zone. She wished it had a clock. Then you would know how long you would have to wait. Five minutes to go? four minutes? Chemical mist drizzled down her faceplate. It was like driving a car in the rain when the windshield wipers are broken; you can't see a thing. Shit, shit, shit, she thought.

At the Institute, there is a Level 4 biocontainment hospital called the Slammer, where a patient can be treated by doctors and nurses wearing space suits. If you are exposed to a hot agent and you go into the Slammer and fail to come out alive, then your body is taken to a nearby Level 4 biocontainment morgue, known as the Submarine. The soldiers around the Institute call the morgue the Submarine because its main door is made of heavy steel and looks like a pressure door in a submarine.

Son of a bitch! she thought. They'll put me into the Slammer. And Tony will be filling out accident reports while I'm breaking with Ebola. And a week later, I'll be in the Submarine. Shit! Jerry's in Texas. And I didn't go to the bank today. There's no money in the house. The kids are home with Mrs. Trapane, *and she needs to be paid.* I didn't go to the market today. There's no food in the house. How are the kids going to eat if I'm in the Slammer? Who's going to stay with them tonight? Shit, shit, shit!

The shower stopped. She opened the door and flung herself into the staging area. She came out of the space suit fast. She shucked it. She leaped out of it. The space suit slapped to the concrete floor, wet, dripping with water.

As her right arm came out of the suit, she saw that the sleeve of her scrub suit was dark wet and her inner glove was red.

That space-suit glove had been a leaker. Ebola blood had run over her innermost glove. It had smeared down on the latex, right against her skin, right against the Band-Aid. Her last glove was thin and translucent, and she could see the Band-Aid through it, directly under the Ebola blood. Her heart pounded, and she almost threw up—her stomach contracted and turned over, and she felt a gag reflex in her throat. The puke factor. It is a sudden urge to throw up when you find yourself unprotected in the presence of a Biosafety Level 4 organism. Her mind raced: What now? I've got an undeconned glove—Ebola blood in here. Oh, *Jesus*. What's the procedure here? What do I have to do now?

Tony Johnson's blue figure moved in the air lock, and she heard the nozzles begin to hiss. He had begun the decon cycle. It would be seven minutes before he could answer any questions.

The main question was whether any blood had penetrated the last glove to the cut. Five or ten Ebola-virus particles suspended in a droplet of blood could easily slip through a pinhole in a surgical glove, and that might be enough to start an explosive infection. This stuff could amplify itself. A pinhole in a glove might not be visible to the eye. She went over to the sink and put her hand under the faucet to rinse off the blood and held it there for a while. The water carried the blood down the drain, where the waste water would be cooked in heated tanks.

Then she pulled off the last glove, holding it delicately by the cuff. Her right hand came out caked with baby powder, her fingernails short, no nail polish, no rings, knuckles scarred by a bite from a goat that had nipped her when she was a child, and a Band-Aid on the palm.

She saw blood mixed with the baby powder.

Please, please, make it *my* blood.

Yes—it was her own blood. She had bled around the edges of the Band-Aid. She did not see any monkey blood on her hand.

She put the last glove under the faucet. The water was running, and it filled up the glove. The glove swelled up like a water balloon. She dreaded the sudden appearance of a thread of water squirting from the glove, the telltale of a leak, a sign that her life was over. The glove fattened and held. No leaks.

Suddenly her legs collapsed. She fell against the cinder-block wall and slid down it, feeling as if she had been punched in the stomach. She came to rest on the hatbox, the biohazard box that someone had been using as a chair. Her legs kicked out, and she went limp and leaned back against the wall. That was how Tony Johnson found her when he emerged from the air lock.

The accident report concluded that Major N. Jaax had not been exposed to Ebola virus. Her last glove had remained intact, and since everyone believed that the agent was transmitted through direct contact with blood and bodily fluids, there did not seem to be any way for it to have entered her bloodstream, even though it had breached her space suit. She drove home that night having escaped the Slammer by the skin of her last glove. She had almost caught Ebola from a dead monkey, who had caught it from a young woman named Mayinga, who had caught it from a nun who had crashed and bled out in the jungles of Zaire in years gone by.

She called Jerry that night in Texas. "Guess what? I had this little problem today. I had a near-Ebola experience." She told him what had happened.

He was appalled. "God-damn it, Nancy! I told you not to get involved with that Ebola virus! That *fucking* Ebola!" And he went into a ten-minute diatribe about the dangers of doing hot work in a space suit, especially with Ebola.

She remained calm and did not argue with him. She realized he wasn't angry with her, just scared. She let Jerry run on, and when he had gotten it all out of his system and was starting to taper off, she told him that she felt confident that everything was going to be all right.

Meanwhile, he was surprised at how calm his wife seemed. He would have flown home that night if he had perceived any inkling of distress in her.

The Ebola experiments were not a success in the sense that the drugs had no effect on the virus. All of Gene Johnson's infected monkeys died no matter what drugs they were given. They *all* died. The virus absolutely nuked the monkeys. It was a complete slate wiper. The only survivors of the experiment were the two control monkeys—the healthy, uninfected monkeys that lived in cages across the room from the sick monkeys. The control monkeys had not been infected with Ebola, and so, as expected, they had not become sick.

Then, two weeks after the incident with the bloody glove, something frightening happened in the Ebola rooms. The two healthy monkeys developed red eyes and bloody noses, and they crashed and bled out. They had never been deliberately infected with Ebola virus, and they had not come near the sick monkeys. They were separated from the sick monkeys by open floor.

If a healthy person were placed on the other side of a room from a person who was sick with AIDS, the AIDS virus would not be able to drift across the room through the air and infect the healthy person. But Ebola had drifted across a room. It had moved quickly, decisively, and by an unknown route. Most likely the control monkeys inhaled it into their lungs. "It got there somehow," Nancy Jaax would say to me as she told me the story some years later. "Monkeys spit and throw stuff. And when the caretakers wash the cages down with water hoses, that can create an aerosol of droplets. It probably traveled through the air in aerosolized secretions. That was when *I* knew that Ebola can travel through the air."

Ebola River

On July 6, 1976, five hundred miles northwest of Mount Elgon, in southern Sudan, near the fingered edge of the central-African rain forest, a man who is known to Ebola hunters as Yu. G. went into shock and died with blood running from the orifices of his body. He is referred to only by his initials. Mr. Yu. G. was the first identified case, the index case, in an outbreak of an unknown virus.

Mr. Yu. G. was a storekeeper in a cotton factory in the town of Nzara. The population of Nzara had grown in recent years—the town had experienced, in its own way, the human population explosion that is occurring throughout the equatorial regions of the earth. The people of that area in southern Sudan are the Zande, a large tribe. The country of the Zande is savanna mixed with riverine forest, beautiful country, where acacia trees cluster along the banks of seasonal rivers. African doves perch in the trees and call their drawn-out calls. The land between the rivers is a sea of elephant grass, which can grow ten feet high. As you head south, toward Zaire, the land rises and forms hills, and the forest begins to spread away from the rivers and thickens into a closed canopy, and you enter the rain forest. The land around the town of Nzara held rich plantations of teak and fruit trees and cotton. People were poor, but they worked hard and raised large families and kept to their tribal traditions.

Mr. Yu. G. was a salaried man. He worked at a desk in a room piled with cotton cloth at the back of the factory. Bats roosted in the ceiling of the room near his desk. If the bats were infected with Ebola, no one has been able to prove it. The virus may have entered the cotton factory by some un-known route—perhaps in insects trapped in the cotton fibers, for example, or in rats that lived in the factory. Or, possibly, the virus had nothing to do with the cotton factory, and Mr. Yu. G. was infected somewhere else. He did not go to a hospital, and died on a cot in his family compound. His family gave him a traditional Zande funeral and left his body under a mound of stones in a clearing of elephant grass. His grave has been visited more than once by doctors from Europe and America, who want to see it and reflect on its meaning, and pay their respects to the index case of what later became known as Ebola Sudan.

He is remembered today as a "quiet, unremarkable man." No photograph was taken of him during his lifetime, and no one seems to remember what he looked like. He wasn't well known, even in his hometown. They say that his brother was tall and slender, so perhaps he was, too. He passed through the gates of life unnoticed by anyone except his family and a few of his co-workers. He might have made no difference except for the fact that he was a host.

His illness began to copy itself. A few days after he died, two other salaried men who worked at desks near him in the same room broke with bleeding, went into shock, and died with massive hemorrhages from the natural openings of the body. One of the dead men was a popular fellow known as P. G. Unlike the quiet Mr. Yu. G., he had a wide circle of friends, including several mistresses. He spread the agent far and wide in the town. The agent jumped easily from person to person, apparently through touching and sexual contact. It was a fast spreader, and it could live easily in people. It passed through as many as sixteen generations of infection as it jumped from person to person in Sudan. It also killed many of its hosts. While this is not necessarily in the best interest of the virus, if the virus is highly contagious, and can jump fast enough from host to host, then it does not matter, really, what happens to the previous host, because the virus can amplify itself for quite a while, at least until it kills off much of the population of hosts. Most of the fatal cases of Ebola Sudan can be traced back through chains of infection to the quiet Mr. Yu. G. A hot strain radiated out of him and nearly devastated the human population of southern Sudan. The strain burned through the town of Nzara and reached eastward to the town of Maridi, where there was a hospital.

It hit the hospital like a bomb. It savaged patients and snaked like chain lightning out from the hospital through patients' families. Apparently the medical staff had been giving patients injections with dirty needles. The virus jumped quickly through the hospital via the needles, and then it hit the medical staff. A characteristic of a lethal, contagious, and incurable virus is that it quickly gets into the medical people. In some cases, the medical system may intensify the outbreak, like a lens that focuses sunlight on a heap of tinder.

The virus transformed the hospital at Maridi into a morgue. As it jumped from bed to bed, killing patients left and right, doctors began to notice signs of mental derangement, psychosis, depersonalization, zombie-like behavior. Some of the dying stripped off their clothes and ran out of the hospital, naked and bleeding, and wandered through the streets of the town, seeking their homes, not seeming to know what had happened or how they had gotten into this condition. There is no doubt that Ebola damages the brain and causes psychotic dementia. It is not easy, however, to separate brain damage from the effects of fear. If you were trapped in a hospital where people were dissolving in their beds, you might try to escape, and if you were a bleeder and frightened, you might take off your clothes, and people might think you had gone mad.

The Sudan strain was more than twice as lethal as Marburg virus—its case-fatality rate was 50 percent. That is, fully half of the people who came down with it ended up dying, and quickly. This was the same kind of fatality rate as was seen with the black plague during the Middle Ages. If the Ebola Sudan virus had managed to spread out of central Africa, it might have entered Khartoum in a few weeks, penetrated Cairo a few weeks after that, and from there it would have hopped to Athens, New York, Paris, London, Singapore—it would have gone everywhere on the planet. Yet that never happened, and the crisis in Sudan passed away unnoticed by the world at large. What happened in Sudan could be compared to the secret detonation of an atomic bomb. If the human race came close to a major biological accident, we never knew it.

For reasons that are not clear, the outbreak subsided, and the virus vanished. The hospital at Maridi had been the epicenter of the emergence. As the virus ravaged the hospital, the surviving medical staff panicked and ran off into the bush. It was probably the wisest thing to do and the best thing that could have happened, because it stopped the use of dirty needles and emptied the hospital, which helped to break the chain of infection.

There was another possible reason why the Ebola Sudan virus vanished. It was exceedingly hot. It killed people so fast that they didn't have much time to infect other people before they died. Furthermore, the virus was not airborne. It was not quite contagious enough to start a full-scale disaster. It traveled in blood, and the bleeding victim did not touch very many other people before dying, and so the virus did not have many chances to jump to a new host. Had people been coughing the virus into the air . . . it would have been a different story. In any case, the Ebola Sudan virus destroyed a few hundred people in central Africa the way a fire consumes a pile of straw—until the blaze burns out at the center and ends in a heap of ash—rather than smoldering around the planet, as AIDS has done, like a fire in a coal mine, impossible to put out. The Ebola virus, in its Sudan incarnation, retreated to the heart of the bush, where undoubtedly it lives to this day, cycling and cycling in some unknown host, able to shift its shape, able to mutate and become a new thing, with the potential to enter the human species in a new form.

Two months after the start of the Sudan emergence—the time was now early September 1976—an even more lethal filovirus emerged five hundred miles to the west, in a district of northern Zaire called Bumba Zone, an area of tropical rain forest populated by scattered villages and drained by the Ebola River. The Ebola Zaire strain was nearly twice as lethal as Ebola Sudan. It seemed to emerge out of the stillness of an implacable force brooding on an inscrutable intention. To this day, the first human case of Ebola Zaire has never been identified.

In the first days of September, some unknown person who probably lived somewhere to the south of the Ebola River perhaps touched something bloody. It might have been monkey meat—people in that area hunt monkeys for food—or it might have been the meat of some other animal, such as an elephant or a bat. Or perhaps the person touched a crushed insect, or perhaps he or she was bitten by a spider. Whatever the original host of the virus, it seems that a blood-to-blood contact in the rain forest enabled the virus to move into the human world. The portal into the human race may well have been a cut on this unknown person's hand.

The virus surfaced in the Yambuku Mission Hospital, an upcountry clinic run by Belgian nuns. The hospital was a collection of corrugated tin roofs and whitewashed concrete walls sitting beside a church in the forest, where bells rang and you heard a sound of hymns and the words of the high mass spoken in Bantu. Next door, people stood in line at the clinic and shivered with malaria while they waited for a nun to give them an injection of medicine that might make them feel better.

The mission in Yambuku also ran a school for children. In late August, a teacher from the school and some friends went on a vacation trip to the northern part of Zaire. They borrowed a Land Rover from the mission to make their journey, and they explored the country as they headed northward, moving slowly along rutted tracks, no doubt getting stuck in the mud from time to time, which is the way things go when you try to drive through Zaire. The track was mostly a footpath enclosed by a canopy of trees, and it was always in shadow, as if they were driving through a tunnel. Eventually they came to the Ebola River and crossed it on a ferry barge and continued northward. Near the Obangui River, they stopped at a roadside market, where the schoolteacher bought some fresh antelope meat. One of his friends bought a freshly killed monkey and put it in the back of the Land Rover. Any of the friends could have handled the monkey or the antelope meat while they were bouncing around in the Land Rover.

They turned back, and when the schoolteacher arrived home, his wife stewed the antelope meat, and everyone in the family ate it. The following morning he felt unwell, and so before he reported to his teaching job at the school, he stopped off at the Yambuku Hospital, on the other side of the church, to get an injection of medicine from the nuns.

At the beginning of each day, the nuns at Yambuku Hospital would lay out five hypodermic syringes on a table, and they would use them to give shots to patients all day long. They were using five needles a day to give injections to hundreds of people in the hospital's outpatient and maternity clinics. The nuns and staff occasionally rinsed the needles in a pan of warm water after an injection, to get the blood off the needle, but more often they proceeded from shot to shot without rinsing the needle, moving from arm to arm, mixing blood with blood. Since Ebola virus is highly infective and since as few as five or ten particles of the virus in a blood-borne contact can start an extreme amplification in a new host, there would have been excellent opportunity for the agent to spread.

A few days after the schoolteacher received his injection, he broke with Ebola Zaire. He was the first known case of Ebola Zaire, but he may well have contracted the virus from a dirty needle during his injection at the hospital, which means that someone else might have previously visited the hospital while sick with Ebola virus and earlier in the day received an injection from the same needle that was then used on the schoolteacher. That unknown person probably stood in line for an injection just ahead of the schoolteacher. That person would have started the Ebola outbreak in Zaire. As in Sudan, the emergence of a life form that could in theory have gone around the earth began with one infected person.

The virus erupted simultaneously in fifty-five villages surrounding the hospital. First it killed people who had received injections, and then it moved through families, killing family members, particularly women, who in Africa prepare the dead for burial. It swept through the Yambuku Hospital's nursing staff, killing most of the nurses, and then it hit the Belgian nuns. The first nun to break with Ebola was a midwife who had delivered a stillborn child. The mother was dying of Ebola and had given the virus to her unborn baby. The fetus had evidently crashed and bled out inside the mother's womb. The woman then aborted spontaneously, and the nun who assisted at this grotesque delivery came away from the experience with blood on her hands. The blood of the mother and fetus was radiantly hot, and the nun must have had a small break or cut on the skin of her hands.

She developed an explosive infection and was dead in five days.

There was a nun at the Yambuku Hospital who is known today as Sister M. E. She became gravely ill with *l'épidémie,* or "the epidemic," as they had begun to call it. A priest at Yambuku decided to try to take her to the city of Kinshasa, the capital of Zaire, in order to get her better medical treatment. He and another nun, named Sister E. R., drove Sister M. E. in a Land Rover to the town of Bumba, a sprawl of cinder blocks and wooden shacks that huddles beside the Congo River. They went to the airfield at Bumba and hired a small plane to fly them to Kinshasa, and when they reached the city, they took Sister M. E. to Ngaliema Hospital, a private hospital run by Swedish nurses, where she was given a room of her own. There she endured her agonals and committed her soul to Christ.

Ebola Zaire attacks every organ and tissue in the human body except skeletal muscle and bone. It is a perfect parasite because it transforms virtually every part of the body into a digested slime of virus particles. The seven mysterious proteins that, assembled together, make up the Ebola-virus particle, work as a relentless machine, a molecular shark, and they consume the body as the virus makes copies of itself. Small blood clots begin to appear in the bloodstream, and the blood thickens and slows, and the clots begin to stick to the walls of blood vessels. This is known as pavementing, because the clots fit together in a mosaic. The mosaic thickens and throws more clots, and the clots drift through the bloodstream into the small capillaries, where they get stuck. This shuts off the blood supply to various parts of the body, causing dead spots to appear in the brain, liver, kidneys, lungs, intestines, testicles, breast tissue (of men as well as women), and all through the skin. The skin develops red spots, called petechiae, which are hemorrhages under the skin. Ebola attacks connective tissue with particular ferocity; it multiplies in collagen, the chief constituent protein of the tissue that holds the organs together. (The seven Ebola proteins somehow chew up the body's structural proteins.) In this way, collagen in the body turns to mush, and the underlayers of the skin die and liquefy. The skin bubbles up into a sea of tiny white blisters mixed with red spots known as a maculopapular rash. This rash has been likened to tapioca pudding. Spontaneous rips appear in the skin, and hemorrhagic blood pours from the rips. The red spots on the skin grow and spread and merge to become huge, spontaneous bruises, and the skin goes soft and pulpy, and can tear off if it is touched with any kind of pressure. Your mouth bleeds, and you bleed around your teeth, and you may have hemorrhages from the salivary glands—literally every opening in the body bleeds, no matter how small. The surface of the tongue turns brilliant red and then sloughs off, and is swallowed or spat out. It is said to be extraordinarily painful to lose the surface of one's tongue. The tongue's skin may be torn off during rushes of the black vomit. The back of the throat and the lining of the windpipe may also slough off, and the dead tissue slides down the windpipe into the lungs or is coughed up with sputum. Your heart bleeds into itself; the heart muscle softens and has hemorrhages into its chambers, and blood squeezes out of the heart muscle as the heart beats, and it floods the chest cavity. The brain becomes clogged with dead blood cells, a condition known as sludging of the brain. Ebola attacks the lining of the eyeball, and the eyeballs may fill up with blood: you may go blind. Droplets of blood stand out on the eyelids: you may weep blood. The blood runs from your eyes down your cheeks and refuses to coagulate. You may have a hemispherical stroke, in which one whole side of the body is paralyzed, which is invariably fatal in a case of Ebola. Even while the body's internal organs are becoming plugged with coagulated blood, the blood that streams out of the body cannot clot; it resembles whey being squeezed out of curds. The blood has been stripped of its clotting factors. If you put the runny Ebola blood in a test tube and look at it, you see that the blood is destroyed. Its red cells are broken and dead. The blood looks as if it has been buzzed in an electric blender.

Ebola kills a great deal of tissue while the host is still alive. It triggers a creeping, spotty necrosis that spreads through all the internal organs. The liver bulges up and turns yellow, begins to liquefy, and then it cracks apart. The cracks run across the liver and deep inside it, and the liver completely dies and goes putrid. The kidneys become jammed with blood clots and dead cells, and cease functioning. As the kidneys fail, the blood becomes toxic with urine. The spleen turns into a single huge, hard blood clot the size of a baseball. The intestines may fill up completely with blood. The lining of the gut dies and sloughs off into the bowels and is defecated along with large amounts of blood. In men, the testicles bloat up and turn black-and-blue, the semen goes hot with Ebola, and the nipples may bleed. In women, the labia turn blue, livid, and protrusive, and there may be massive vaginal bleeding. The virus is a catastrophe for a pregnant woman: the child is aborted spontaneously and is usually infected with Ebola virus, born with red eyes and a bloody nose.

Ebola destroys the brain more thoroughly than does Marburg, and Ebola victims often go into epileptic convulsions during the final stage. The convulsions are generalized grand mal seizures—the whole body twitches and shakes, the arms and legs thrash around, and the eyes, sometimes bloody, roll up into the head. The tremors and convulsions of the patient may smear or splatter blood around. Possibly this epileptic splashing of blood is one of Ebola's strategies for success—it makes the victim go into a flurry of seizures as he dies, spreading blood all over the place, thus giving the virus a chance to jump to a new host—a kind of transmission through smearing.

Ebola (and Marburg) multiplies so rapidly and powerfully that the body's infected cells become crystal-like blocks of packed virus particles. These crystals are broods of virus getting ready to hatch from the cell. They are known as bricks. The bricks, or crystals, first appear near the center of the cell and then migrate toward the surface. As a crystal reaches a cell wall, it disintegrates into hundreds of individual virus particles, and the broodlings push through the cell wall like hair and float away in the bloodstream of the host. The hatched Ebola particles cling to cells everywhere in the body, and get inside them, and continue to multiply. It keeps on multiplying until areas of tissue all through the body are filled with crystalloids, which hatch, and more Ebola particles drift into the bloodstream, and the amplification continues inexorably until a droplet of the host's blood can contain a hundred million individual virus particles.

After death, the cadaver suddenly deteriorates: the internal organs, having been dead or partially dead for days, have already begun to dissolve, and a sort of shock-related meltdown occurs. The corpse's connective tissue, skin, and organs, already peppered with dead spots, heated by fever, and damaged by shock, begin to liquefy, and the fluids that leak from the cadaver are saturated with Ebola-virus particles.

When it was all over, the floor, chair, and walls in Sister M. E.'s hospital room were stained with blood. Someone who saw the room told me that after they took her body away for burial (wrapped in many sheets), no one at the hospital could bear to go into the room to clean it up. The nurses and doctors didn't want to touch the blood on the walls and were frankly fearful of breathing the air in the room, too. So the room was closed and locked, and remained that way for days. The appearance of the nun's hospital room after her death may have raised in some minds one or two questions about the nature of the Supreme Being, or, for persons not inclined to theology, the blood on the walls may have served as a reminder of the nature of Nature.

No one knew what had killed the nun, but clearly it was a replicating agent, and the signs and symptoms of the disease were not easy to consider with a calm mind. What also did not lead to calm thoughts were rumors coming out of the jungle to the effect that the agent was wiping out whole villages upriver on the Congo. These rumors were not true. The virus was hitting families selectively, but no one understood this because the flow of news coming from upriver was being choked off. Doctors at the hospital in Kinshasa examined the nun's case and began to suspect that she might have died of Marburg or a Marburg-like agent.

Then Sister E. R., the nun who had traveled with Sister M. E. during the drive to Bumba and the plane flight to Kinshasa, broke with *l'épidémie*. They put her in a private room at the hospital, where she began to die with the same signs and symptoms that had preceded Sister M. E.'s death.

There was a young nurse at the Ngaliema Hospital named Mayinga N. (Her first name was Mayinga and her last name is given as N.) Nurse Mayinga had been caring for Sister M. E. when the nun had died in the bloodstained room. She may have been splattered with the nun's blood or with black vomit. At any rate, Nurse Mayinga developed a headache and fatigue. She knew she was becoming sick, but she did not want to admit to herself what it was. She came from a poor but ambitious family, and she had received a scholarship to go to college in Europe. What worried her was the possibility that if she became ill, she would not be allowed to travel abroad. When the headache came upon her, she left her job at the hospital and disappeared. She dropped out of sight for two days. During that time, she went into the city, hoping to get her travel permits arranged before she became visibly sick. On the first day of her disappearance—the date was October 12, 1976—she spent a day waiting in lines at the offices of the Zairean foreign ministry, trying to get her papers straight.

The next day, October 13, she felt worse, but instead of reporting to work, again she went into the city. This time, she took a taxi to the largest hospital in Kinshasa, the Mama Yemo Hospital. By now, as her headache became blinding and her stomach pain increased, she must have been terribly frightened. Why didn't she go to the Ngaliema Hospital to seek treatment where she worked and where the doctors would have taken care of her? It must have been a case of psychological denial. She did not want to admit, even to herself, that she had been infected. Perhaps she had a touch of malaria, she hoped. So she went to Mama Yemo Hospital, the hospital of last resort for the city's poor, and spent hours waiting in a casualty ward jammed with ragged people and children.

I can see her in my mind's eye—Nurse Mayinga, the source of the virus in the United States Army's freezers. She was a pleasant, quiet, beautiful young African woman, about twenty years old, in the prime of her life, with a future and dreams, hoping somehow that what was happening to her could not be happening. They say that her parents loved her dearly, that she was the apple of their eye. Now she is sitting in the casualty ward at Mama Yemo among the cases of malaria, among the large-bellied children in rags, and no one is paying any attention to her because all she has is a headache and red eyes. Perhaps she has been crying, perhaps that is why her eyes are red. A doctor gives her a shot for malaria and tells her that she should be in quarantine for her illness. But there is no room in the quarantine ward at Mama Yemo Hospital; so she leaves the hospital and hails another taxi. She tells the driver to take her to another hospital, to University Hospital, where perhaps the doctors can treat her. But when she arrives at University Hospital, the doctors can't seem to find anything wrong with her, except for possibly some signs of malaria. Her headache is getting worse. She is sitting in the waiting room of this hospital, and as I try to imagine her there, I am almost certain she is crying. Finally she does the only thing left for her to do. She returns to Ngaliema Hospital and asks to be admitted as a patient. They put her in a private room, and there she falls into lethargy, and her face freezes into a mask.

News of the virus and what it did to people had been trickling out of the forest, and now a rumor that a sick nurse had wandered around Kinshasa for two days, having face-to-face contact with many people in crowded rooms and public places, caused a panic in the city. The news spread first along the mission grapevine and through government employees and among the diplomats at cocktail parties, and finally the rumors began to reach Europe. When the story reached the offices of the World Health Organization in Geneva, the place went into a full-scale alert. People who were there at the time said that you could feel fear in the hallways, and the director looked like a visibly shaken man. Nurse Mayinga seemed to be a vector for an explosive chain of lethal transmission in a crowded third-world city with a population of two million people. Officials at the WHO began to fear that Nurse Mayinga would become the vector for a world-wide plague. European governments contemplated blocking flights from Kinshasa. The fact that one infected person had wandered around the city for two days when she should have been isolated in a hospital room began to look like a species-threatening event.

President Mobutu Sese Seko, the maximum leader of Zaire, sent his army into action. He stationed soldiers around Ngaliema Hospital with orders to let no one enter or leave except doctors. Much of the medical staff was now under quarantine inside the hospital, but the soldiers made sure that the quarantine was enforced. President Mobutu also ordered army units to seal off Bumba Zone with roadblocks and to shoot anyone trying to come out. Bumba's main link with the outside world was the Congo River. Captains of riverboats had heard about the virus by this time, and they refused to stop their boats anywhere along the length of the river in Bumba, even though people beseeched them from the banks. Then all radio contact with Bumba was lost. No one knew what was happening upriver, who was dying, what the virus was doing. Bumba had dropped off the face of the earth into the silent heart of darkness.

As the first nun at Ngaliema Hospital, Sister M. E., lay dying, her doctors had decided to give her a so-called agonal biopsy. This is a rapid sampling of tissue, done close to the moment of death instead of a full autopsy. She was a member of a religious order that prohibited autopsies, but the doctors very much wanted to know what was replicating inside her. As the terminal shock and convulsions came over her, they inserted a needle into her upper abdomen and sucked out a quantity of liver. Her liver had begun to liquefy, and the needle was large. A fair amount of the nun's liver traveled up the needle and filled a biopsy syringe. Possibly it was during this agonal biopsy that her blood squirted on the walls. The doctors also took some samples of blood from her arm and put it in glass tubes. The nun's blood was precious beyond measure, since it contained the unknown hot agent.

The blood was flown to a national laboratory in Belgium and to the English national laboratory, the Microbiological Research Establishment at Porton Down, in Wiltshire. Scientists at both labs began racing to identify the agent. Meanwhile, at the Centers for Disease Control in Atlanta, Georgia —the C.D.C.—scientists were feeling left out, and were still scrambling to get their hands on a little bit of the nun's blood, making telephone calls to Africa and Europe, pleading for samples.

There is a branch of the C.D.C. that deals with unknown emerging viruses. It's called the Special Pathogens Branch. In 1976, at the time of the Zaire outbreak, the branch was being run by a doctor named Karl M. Johnson, a virus hunter whose home terrain had been the rain forests of Central and South America. (He is not related to Gene Johnson, the civilian virus hunter, or to Lieutenant Colonel Tony Johnson, the pathologist.) Karl Johnson and his C.D.C. colleagues had heard almost nothing about the occurrences upriver in Zaire —all they knew was that people in Zaire were dying of a "fever" that had "generalized symptoms"—no details had come in from the bush or from the hospital where the nun had just died. Yet it sounded like a bad one. Johnson telephoned a friend of his at the English lab, in Porton Down, and reportedly said to him, "If you've got any little dregs to spare of that nun's blood, we'd like to have a look at it." The Englishman agreed to send it to him, and what he received was literally dregs.

The nun's blood arrived at the C.D.C. in glass tubes in a box lined with dry ice. The tubes had cracked and broken during shipment, and raw, rotten blood had run around inside the box. A C.D.C. virologist named Patricia Webb— who was then married to Johnson—opened the box. She found that the package was sticky with blood. The blood looked like tar. It was black and gooey, like Turkish coffee. She put on rubber gloves, but other than that, she did not take any special precautions in handling the blood. Using some cotton balls, she managed to dab up some of the black stuff, and then by squeezing the cotton between her gloved fingertips, she collected a few droplets of it, just enough to begin testing it for viruses.

Patricia Webb put some of the black blood droplets into flasks of monkey cells, and pretty soon the cells got sick and began to die—they burst. The unknown agent could infect monkey cells and pop them.

Another C.D.C. doctor who worked on the unknown virus was Frederick A. Murphy, a virologist who had helped to identify Marburg virus. He was and is one of the world's leading electron-microscope photographers of viruses. (His photographs of viruses have been exhibited in art museums.) Murphy wanted to take a close look at those dying cells to see if he could photograph a virus in them. On October 13— the same day Nurse Mayinga was sitting in the waiting rooms of hospitals in Kinshasa—he placed a droplet of fluid from the cells on a small screen and let it dry, and he put it in his electron microscope to see what he could see.

He couldn't believe his eyes. The sample was jammed with virus particles. The dried fluid was shot through with something that looked like string. His breath stopped in his throat. He thought, *Marburg*. He believed he was looking at Marburg virus.

Murphy stood up abruptly, feeling strange. That lab where he had prepared these samples—that lab was hot. That lab was as hot as hell. He went out of the microscope room, closing the door behind him, and hurried down a hallway to the laboratory where he had worked with the material. He got a bottle of Clorox bleach and scrubbed the room from top to bottom, washing countertops and sinks, everything, with bleach. He really scoured the place. After he had finished, he found Patricia Webb and told her what he had seen in his microscope. She telephoned her husband and said to him, "Karl, you'd better come quick to the lab. Fred has looked at a specimen, and he's got *worms*."

Staring at the worms, they tried to classify the shapes. They saw snakes, pigtails, branchy, forked things that looked like the letter *Y,* and they noticed squiggles like a small *g,* and bends like the letter *U,* and loopy 6s. They also noticed a classic shape, which they began calling the shepherd's crook. Other Ebola experts have taken to calling this loop the eyebolt, after a bolt of the same name that can be found in a hardware store. It could also be described as a Cheerio with a long tail.

The next day, Patricia Webb ran some tests on the virus and found that it did not react to any of the tests for Marburg or any other known virus. Therefore, it was an unknown

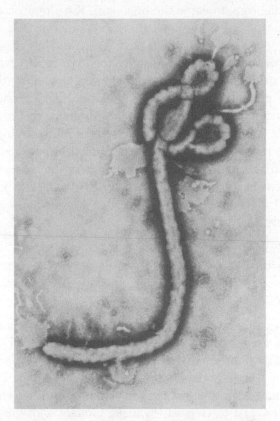

A single Ebola-virus particle with a pronounced "shepherd's crook"—in this case, a tangled double crook. This is one of the first photographs ever made of Ebola. It was taken on October 13, 1976, by Frederick A. Murphy, then of the Centers for Disease Control—on the same day that Nurse Mayinga wandered around Kinshasa. The lumpy ropelike braided features in the particle are the mysterious structural proteins. They surround a single strand of RNA, which is the virus's genetic code. Magnification: 112,000 times.

agent, a new virus. She and her colleagues had isolated the strain and shown that it was something new. They had earned the right to name the organism. Karl Johnson named it Ebola.

Karl Johnson has since left the C.D.C., and he now spends a great deal of his time fly-fishing for trout in Montana. He does consulting work on various matters, including the design of pressurized hot zones. I learned that he could be reached at a fax number in Big Sky, Montana, so I sent him a fax. In it, I said that I was fascinated by Ebola virus. My fax was received, but there was no reply. So I waited a day and then sent him another fax. It fell away into silence. The man must have been too busy fishing to bother to answer. After I had given up hope, my fax machine suddenly emitted this reply:

> Mr. Preston:
>   Unless you include the feeling generated by gazing into the eyes of a waving confrontational cobra, "fascination" is not what I feel about Ebola. How about shit scared?

Two days after he and his colleagues isolated Ebola virus for the first time, Karl Johnson headed for Africa in the company of two other C.D.C. doctors, along with seventeen boxes of gear, to try to organize an effort to stop the virus in Zaire and Sudan (the outbreak in Sudan was still going on). They flew first to Geneva, to make contact with the World Health Organization, where they found that the WHO knew very little about the outbreaks. So the C.D.C. doctors organized their equipment and packed more boxes, getting ready to go to the Geneva airport, from where they would fly to Africa. But then, at the last possible moment, one of C.D.C. doctors panicked. It is said he was the doctor assigned to go to Sudan, and it is said he was afraid to proceed any farther. It was not an unusual situation. As Karl Johnson explained to me, "I've seen young physicians run from these hemorrhagic viruses, literally. They're unable to work in the middle of an outbreak. They refuse to get off the plane.''

Johnson, one of the discoverers of Ebola virus, preferred to talk about these events while fly-fishing. (''We've got to keep our priorities straight,'' he explained to me.) So I flew to Montana and spent a couple of days with him fishing for brown trout on the Bighorn River. It was October, the weather had turned clear and warm, and the cottonwood trees along the banks were yellow and rattled in a south wind. Standing waist-deep in a mutable slick of the river, wearing sunglasses, with a cigarette hanging from the corner of his mouth and a fly rod in his hand, Johnson ripped his line off the water and laid a cast upstream. He was a lean, bearded man, with a soft voice that one had to listen for in the wind. He is a great figure in the history of virus hunting, having discovered and named some of the most dangerous life forms on the planet. ''I'm so *glad* nature is not benign,'' he remarked. He studied the water, took a step downstream, and placed another cast. ''But on a day like today, we can pretend nature is benign. All monsters and beasts have their benign moments.''

''What happened in Zaire?'' I asked.

''When we got to Kinshasa, the place was an absolute madhouse,'' he said. ''There was no news coming out of Bumba, no radio contact. We knew it was bad in there, and we knew we were dealing with something new. We didn't know if the virus could be spread by droplets in the air, somewhat like influenza. If Ebola *had* spread easily through the air, the world would be a very different place today.''

''How so?''

''There would be a lot fewer of us. It would have been *exceedingly* difficult to contain that virus if it had had any major respiratory component. I did figure that if Ebola was the Andromeda strain—incredibly lethal and spread by droplet infection—there wasn't going to be any safe place in the world anyway. It was better to be working at the epicenter than to get the infection at the London opera.''

"Are you worried about a species-threatening event?''

He stared at me. "What the hell do you mean by that?''

"I mean a virus that wipes us out.''

"Well, I think it could happen. Certainly it hasn't happened yet. I'm not worried. More likely it would be a virus that reduces us by some percentage. By thirty percent. By ninety percent.''

"Nine out of ten humans killed? And you're not bothered.''

A look of mysterious thoughtfulness crossed his face. "A virus can be useful to a species by thinning it out,'' he said.

A scream cut the air. It sounded nonhuman.

He took his eyes off the water and looked around. "Hear that pheasant? That's what I like about the Bighorn River,'' he said.

"Do you find viruses beautiful?''

"Oh, yeah,'' he said softly. "Isn't it true that if you stare into the eyes of a cobra, the fear has another side to it? The fear is lessened as you begin to see the essence of the beauty. Looking at Ebola under an electron microscope is like looking at a gorgeously wrought ice castle. The thing is so cold. So totally pure.'' He laid a perfect cast on the water, and eddies took the fly down.

Karl Johnson became the chief of an international WHO team that gathered in Kinshasa to try to stop the Ebola outbreak.

The other C.D.C. doctor, Joel Breman, who had flown with Johnson to Zaire, became a member of a field exploration team that boarded an aircraft bound for the interior to see what was going on in Bumba. The airplane was a C-130 Buffalo troop-transport, an American-made military aircraft that belonged to the Zairean Air Force. It happened to be President Mobutu's private plane, equipped with leopard-skin seats, folding beds, and a wet bar, a sort of flying presidential palace that ordinarily took the president and his family on vacations to Switzerland, but now it carried the WHO team into the hot zone, following the Congo River north by east. They sat on the leopard-skin seats and stared out the windows at endless tracts of rain forest and brown river, a featureless blanket broken by the occasional gleam of an oxbow lake or a cluster of round huts strung like beads on a barely visible road or footpath. As he leaned against the window and watched the terrain unfold into the heart of Africa, Breman became terrified of coming to earth. It was safe in the air, high above the immeasurable forest, but down there . . . It began to dawn on him that he was going to Bumba to die. He had recently been assigned to Michigan as a state epidemiologist, and suddenly he had been called to Africa. He had left his wife back home in Michigan with their two children, and he began to suspect that he would never see them again. He had brought an overnight bag with a toothbrush, and he had managed to pack a few paper surgical masks and some gowns and rubber gloves into the bag. He did not have proper equipment for handling a hot agent. The Buffalo descended, and the town of Bumba appeared, a rotting tropical port spread out along the Congo River.

The Buffalo landed at an airstrip outside the town. The plane's Zairean crew was terrified, afraid to breathe the air, and they left the propellers idling while they shoved the doctors down the gangway and heaved their bags out after them. The doctors found themselves standing in the backwash of the Buffalo as it accelerated to take off.

In the town, they met with the governor of Bumba Zone. He was a local politician, quite distraught. He had found himself in deep waters, in over his head. "We are in a bad way," he told the doctors. "We have not been able to get salt or sugar." His voice trembled on the edge of weeping as he added, "We have not even been able to get beer."

A Belgian doctor on the team knew how to handle this

situation. With a dramatic flourish, he placed a black airline pilot's bag on the table. Then he turned the bag upside down, and wads of currency slapped out, making an impressive heap. "Governor, perhaps this will make things a little better," he said.

"What are you doing?" Breman said to the Belgian.

The Belgian shrugged and replied in a low voice, "Look, this is the way things are done here."

The governor scooped up the money and pledged his full co-operation together with all the extensive resources of government at his disposal—and he loaned them two Land Rovers.

They pushed north toward the Ebola River.

It was the rainy season, and the "road" was a string of mudholes cut by running streams. Engines howling, wheels spinning, they proceeded through the forest at the pace of a walk, in continual rain and oppressive heat. Occasionally they came to villages, and at each village they encountered a roadblock of fallen trees. Having had centuries of experience with the smallpox virus, the village elders had instituted their own methods for controlling the virus, according to their received wisdom, which was to cut their villages off from the world, to protect their people from a raging plague. It was reverse quarantine, an ancient practice in Africa, where a village bars itself from strangers during a time of disease, and drives away outsiders who appear.

"Who are you? What are you doing?" they shouted to the Land Rovers from behind a barrier of trees.

"We are doctors! We are coming to help!"

Eventually the people would clear away the trees, and the team would proceed deeper into the forest. In a long and desperate day of travel, they penetrated fifty miles away from the Congo River, and finally, toward evening, they came to a row of round, thatched African houses. Beyond the houses stood a white church in the middle of the forest. Around the church, there were two soccer fields, and in the middle of one field they noticed a heap of burned mattresses. Two hundred yards farther on, they came to the Yambuku Mission Hospital, a complex of low, whitewashed buildings made of concrete, with corrugated tin roofs.

The place was as silent as a tomb and appeared to be deserted. The beds were iron or wooden frames without mattresses—the blood-soaked mattresses had been burned in the soccer field—and the floors were clean, spotless, rinsed. The team discovered three surviving nuns and one priest, along with a few devoted African nurses. They had cleaned up the mess after the virus had wiped out everyone else, and now they were busy fogging the rooms with insecticide, in the hope that it might somehow disperse the virus. One room in the hospital had not been cleaned up. No one, not even the nuns, had had the courage to enter the obstetric ward. When Joel Breman and the team went in, they found basins of foul water standing among discarded, bloodstained syringes. The room had been abandoned in the middle of childbirths, where dying mothers had aborted fetuses infected with Ebola. The team had discovered the red chamber of the virus queen at the end of the earth, where the life form had amplified through mothers and their unborn children.

The rains continued all day and night. Around the hospital and the church stood the beautiful ferocious trees, a complex of camphors and teaks. Their crowns entwined and crisscrossed and whispered with rain, and bowed and shifted as troops of monkeys passed through them like currents of wind, leaping from crown to crown, crying their untranslatable cries. The next day, the doctors set out deeper into the forest in their Land Rovers, and they made contact with infected villages, where they found people dying in huts. Some of the victims had been put into isolation huts on the edge of the village—an old African technique for dealing with smallpox. Some of the huts where people had died had been burned down. Already the virus seemed to be petering out, and most of the people who were going to die were already dead, the virus having echoed so swiftly through Bumba. A wave of emotion rolled over Joel Breman as he realized, with the clarity of a doctor who suddenly sees into the heart of things, that the victims had received the infection from the hospital. The virus had taken root with the nuns and had done its work among those who had sought help from them. In one village, he examined a man dying of Ebola. The man sat in a chair, holding his stomach and leaning forward with pain, and blood streamed around his teeth.

They tried to reach Kinshasa by radio, to tell Karl Johnson and the others that the epidemic had already peaked. A week later, they were still trying to make radio contact, but they could not get through. They traveled back to the town of Bumba and waited by the river. One day an airplane droned overhead. It circled the town once and touched down, and they ran for it.

At the Ngaliema Hospital in Kinshasa, Nurse Mayinga had been put into a private room, which was accessible through a kind of entry room, a gray zone, where the nurses and staff were supposed to put on bioprotective gear before they entered. Mayinga was cared for by a South African doctor named Margaretha Isaäcson, who at first wore a military gas mask, but it became increasingly uncomfortable in the tropical heat. She thought to herself, I can't bear it, I'll be surprised if I come out of this alive anyway. That made her think about her own children. She thought, My children are grown up, they are no great responsibility. And she removed her mask and treated the dying girl face-to-face.

Dr. Isaäcson did everything she could to save Mayinga, but she was as helpless before the agent as medieval doctors had been in the face of the black plague. ("This was not like AIDS," she later recalled to me. "AIDS is child's play compared with this.") She gave Nurse Mayinga ice cubes to suck on, which helped to ease the pain in her throat, and she gave her Valium to try to stave off her apprehension of what lay ahead.

"I know I am dying," Mayinga said to her.

"That's nonsense. You are not going to die," Dr. Isaäcson replied.

When Mayinga's bleeding began, it came from her mouth and nose. It never came in a rush, but the blood dripped and ran and would not stop and would not clot. It was a hemorrhagic nosebleed, the kind that does not stop until the heart stops beating. Eventually Dr. Isaäcson gave her three transfusions of whole blood to replace what she lost in nosebleeds. Mayinga remained conscious and despondent until the end. In the final stage, her heart developed a galloping beat. Ebola had entered her heart. Mayinga could feel her heart going blubbery inside her chest as Ebola worked its way through her heart, and it frightened her unspeakably. That night, she died of a heart attack.

Her room was contaminated with blood, and there was also the question of the two nuns' rooms, both of which were still locked and bloodstained. Dr. Isaäcson said to the staff, "I won't be of much use to you now," and she got a bucket and mop and washed the rooms.

Medical teams fanned out into Kinshasa and managed to locate thirty-seven people who had had face-to-face contact with Mayinga during the time when she had wandered around the city. They set up two biocontainment pavilions at the hospital and shut the people up for a couple of weeks. They wrapped the cadavers of the nuns and Nurse Mayinga in sheets soaked in chemicals, then double-bagged the mummies in plastic and put each one in an airtight coffin with a screw-down lid, and held the funeral services at the hospital, under the watch of doctors.

Karl Johnson, having heard nothing from the team of doctors upriver in Bumba, wondered if they were dead, and assumed that the virus was about to go on a burn through the city. He organized a floating hospital ship and had it moored in the Congo River. It was an isolation ship for doctors. The city would be the hot zone, and the floating ship would be the gray area, the place of refuge for the doctors. Approximately a thousand Americans were living in Zaire at the time. In the United States, the Army's Eighty-second Airborne Division went on alert and prepared to evacuate the Americans by air as soon as the first Ebola cases started popping up in the city. But to the strange and wonderful relief of Zaire and the world, the virus never went on a burn through the city. It subsided on the headwaters of the Ebola River and went back to its hiding place in the forest. The Ebola agent seemed not to be contagious in face-to-face contacts. It did not seem to be able to travel through the air. No one caught the virus from Nurse Mayinga, even though she had been in close contact with at least thirty-seven people. She had shared a bottle of soda pop with someone, and not even that person became ill. The crisis passed.

Cardinal

1987 SEPTEMBER

As with Ebola, the secret hiding place of the Marburg agent was unknown. After erupting in Charles Monet and Dr. Shem Musoke, Marburg dropped out of sight, and no one could say where it had gone. It seemed to vanish off the face of the earth, but viruses never go away, they only hide, and Marburg continued to cycle in some reservoir of animals or insects in Africa.

On the second day of September 1987, around supper-time, Eugene Johnson, the civilian biohazard expert attached to USAMRIID, stood in a passenger-arrival area outside the customs gates at Dulles International Airport, near Washington. He was waiting for a KLM flight from Amsterdam, which carried a passenger who had come from Kenya. A man with a duffle bag passed through customs, and he and Johnson nodded to each other. ("I'm going to leave this person's name out of it. Let's just say he was someone I knew, someone I trust," Johnson explained to me.) The man laid down the duffle bag at Johnson's feet, unzipped it, and pulled out a wad of bath towels wrapped around something. Pulling off the towels, he revealed an unmarked cardboard box wound with tape. He handed the box to Johnson. They had little to say to each other. Johnson carried the box out of the terminal building, put it in the trunk of his car, and drove to the Institute. The box held blood serum from a ten-year-old Danish boy who will be called Peter Cardinal. He had died a day or so earlier at Nairobi Hospital with a combination of extreme symptoms that suggested an unidentified Level 4 virus.

As he drove to the Institute, Johnson wondered just what he was going to do with the box. He was inclined to sterilize its contents in an oven and then incinerate it. Just cook it and burn it, and forget it. Most of the samples that came into the Institute—and samples of blood and tissue arrived constantly from all parts of the world—contained nothing unusual, no interesting viruses. In other words, most of the samples were false alarms. Johnson wasn't sure he wanted to take the time to analyze this boy's blood serum, if, in all probability, nothing would be found in it. By the time he pulled into the gates of Fort Detrick, he had decided to go ahead. He knew the work would keep him up most of the night, but it had to be done immediately, before the blood serum deteriorated.

Johnson put on a surgical scrub suit and rubber gloves, and carried the box into the Level 3 staging area of the Ebola suite, where he opened the box, revealing a mass of foam peanuts. Out of the peanuts he fished a metal cylinder sealed with tape and marked with a biohazard symbol. Along the wall of the staging area was a row of stainless-steel cabinets with rubber gloves protruding into them. They were Biosafety Level 4 cabinets. They could be sealed off from the outside world while you handled a hot agent inside them with the rubber gloves. These cabinets were similar in design to the safety cabinets that are used for handling nuclear-bomb parts. Here the cabinets were designed to keep human beings from coming into direct contact with Nature. Johnson unscrewed some wingnuts and opened a door in the cabinets, and placed the metal cylinder inside. He closed and tightened the door.

Next, he put his hands into the gloves, picked up the cylinder and, looking through a window to see what he was doing, peeled the tape off the cylinder. The tape stuck to his rubber gloves, and he couldn't get it off. Damn! he swore to himself. It was now eight o'clock at night, and he would never get home. Finally he got the cylinder open. Inside it was a wad of paper towels soaked in bleach. He pulled apart the wad and found a Ziploc bag. It contained a couple of plastic tubes with screw tops. He unscrewed them and shook out two very small plastic vials containing golden liquid: Peter Cardinal's blood serum.

The boy's mother and father worked for a Danish relief organization in Kenya, and lived in a house in the town of Kisumu, on Lake Victoria. Peter had been a student at a boarding school in Denmark. That August, a few weeks before he died, he had gone to Africa to visit his parents and his older sister. She was a student at a private school in Nairobi. She and Peter were very close, and while Peter was visiting his family in Kenya, the two young people spent most of their time together—brother and sister, best friends.

The Cardinal family went on vacation after Peter's arrival, and traveled by car through Kenya—his mother and father wanted to show him the beauty and sweetness of Africa. They were visiting Mombasa, staying in a hotel by the sea, when Peter developed red eyes. His parents took him to the hospital, where the doctors examined him and concluded that he had come down with malaria. His mother did not believe it was malaria. She began to perceive that her son was dying, and she became frantic. She insisted that he be taken to Nairobi for treatment. The Flying Doctors, an air-ambulance service, picked him up, and he was flown to Nairobi and rushed to Nairobi Hospital, where he came under the care of Dr. David Silverstein, who had also taken care of Dr. Musoke after Charles Monet had spewed the black vomit into Musoke's eyes.

• • •

"Peter Cardinal was a blond-haired, blue-eyed guy, a tall, thin guy, a fit-looking ten-year-old," Dr. Silverstein recalled as we drank coffee and tea at a table in the shopping mall near his house outside Washington. A small girl sitting nearby burst into wails, and her mother hushed her. Crowds of shoppers passed by our table. I kept my eyes on Dr. Silverstein's face—steel-rimmed glasses, mustache, eyes that gazed into space—as he recalled the unusual death he had seen, which he spoke of in a matter-of-fact way. "When Peter came to me, he was febrile, but he was very with it, very alert and communicative. We gave him an X ray. His lungs were fluffy." A kind of watery mucus had begun to collect in the boy's lungs, which caused him difficulty in breathing. "It was a typical ARDS picture—acute respiratory distress syndrome—like early pneumonia," Dr. Silverstein said. "Shortly afterward, he started turning bluish on me. He had blue in his fingertips. Also, he had little red spots. I had everybody glove up before they handled him. We suspected he had Marburg, but we didn't have the paranoia we had had with Dr. Musoke. We just took precautions. In twenty-four hours, he was on a respirator. We noted that he bled easily from puncture sites, and he had deranged liver functions. The small red spots became large, spontaneous bruises. He turned black-and-blue. Then his pupils dilated up on us. That was a sign of brain death. He was bleeding around the brain."

The boy swelled up, and his skin filled with pockets of blood. In some places, the skin almost separated from the underlying tissue. This happened during the last phase, while he was on the respirator. It is called third spacing. If you bleed into the first space, you bleed into your lungs. If you bleed into the second space, you bleed into your stomach and intestines. If you bleed into the third space, you bleed into the space between the skin and the flesh. The skin puffs up and separates from the flesh like a bag. Peter Cardinal had bled out under his skin.

The more one contemplates the hot viruses, the less they look like parasites and the more they begin to look like predators. It is a characteristic of a predator to become invisible to its prey during the quiet and sometimes lengthy stalk that precedes an explosive attack. The savanna grass ripples on the plains, and the only sound in the air is the sound of African doves calling from acacia trees, a pulse that goes on through the heat of the day and never slows and never ends. In the distance, in the flickering heat, in the immense distance, a herd of zebras graze. Suddenly from the grass comes a streak of movement, and a lion is among them and hangs on a zebra's throat. The zebra gives out a barking cry, choked off, and the two interlocked beings, the predator and the prey, spin around in a dance, until you lose sight of the action in a billow of dust, and the next day the bones have a surface of flies. Some of the predators that feed on humans have lived on the earth for a long time, far longer than the human race, and their origins go back, it seems, almost to the formation of the planet. When a human being is fed upon and consumed by one of them, especially in Africa, the event is telescoped against horizons of space and time, and takes on a feeling of immense antiquity.

Peter Cardinal's parents and sister were stunned as they watched him being slowly torn apart by an invisible predator. They could not comprehend his suffering or reach him to give him comfort. As the blood poured into his third space, his eyes remained open and dilated, staring, bloody, deep, dark, and bottomless. They didn't know if he could see them, and they couldn't tell what he saw or thought or felt behind the open eyes. The machines hooked up to his scalp were showing flatlines in his brain. There was very little electrical activity in his brain, but now and then the flatlines gave a spooky twitch, as if something continued to struggle inside the boy, some destroyed fragment of his soul.

They had to make a decision about whether to turn off the respirator. Dr. Silverstein said to them, "We're much better off not letting him survive, because of brain death."

"If they had only brought him in sooner from Mombasa," the mother said.

"I'm sorry, but that would not have helped. There was nothing that anyone could have done," Silverstein replied to her. "He was doomed from the beginning."

Working with his hands in the rubber gloves protruding into the cabinets, Gene Johnson took a little bit of the boy's blood serum and dropped it into flasks that contained living cells from a monkey. If anything lived in Peter Cardinal's blood, it might begin to replicate in the monkey cells. Then Johnson went home to get some sleep. The procedure had taken him until three o'clock in the morning to finish.

In the following days, Johnson watched the flasks to see if there were any changes in the monkey cells. He saw that they were bursting and dying. They were infected with something. The Cardinal strain was definitely a hot agent—it killed the cells in vast numbers, and it killed them fast.

Now for the next stage of the virus isolation. He drew off a little bit of fluid from the flasks and injected it into three rhesus monkeys, to infect them with the Cardinal agent. Two of the monkeys died and the third animal went into borderline shock, but somehow pulled through and survived. So the Cardinal agent was viciously hot, a fast replicator, and it could kill monkeys. "I knew goddamned well we had Marburg," Johnson would later say to me.

He took some of the Cardinal strain and injected it into guinea pigs to see if it would infect them. It killed them like flies. Not only that, the testicles of the males swelled up to the size of golf balls and turned purple. The Cardinal strain was a sophisticated organism that knew what it wanted. It could multiply in many different kinds of meat. It was an invasive life form, devastating and promiscuous. It showed a kind of obscenity you see only in nature, an obscenity so extreme that it dissolves imperceptibly into beauty. It made a living somewhere in Africa. What made it particularly interesting was that it multiplied easily in various species, in monkeys, humans, guinea pigs. It was extremely lethal in these species, which meant that its original host was probably not monkeys, humans, or guinea pigs but some other animal or insect that it did not kill. A virus does not generally kill its natural host. The Marburg virus was a traveler: it could jump species; it could break through the lines that separate one species from another, and when it jumped into another species, it had a potential to devastate the species. It did not know boundaries. It did not know what humans are; or perhaps you could say that it knew only too well what humans are: it knew that humans are meat.

As soon as he isolated the Cardinal strain and confirmed that it was Marburg, Johnson turned his attention to the question of where and how Peter Cardinal might have become infected. Where had that kid been? What had he been doing to get himself infected? Exactly where had he traveled? These questions haunted Johnson. He had been trying to find the secret reservoirs of the thread viruses for years.

He telephoned a friend and colleague in Kenya named Dr. Peter Tukei, who was a scientist at the Kenya Medical Research Institute in Nairobi. "We know this is Marburg," Gene said to him. "Can you get a history of the kid? Find out where he was and what he did?"

Dr. Tukei said he would locate the parents and interview them.

A week later, Gene's telephone rang. It was Dr. Tukei on the line. "You know where that kid was?" he said. "He was in Kitum Cave on Mount Elgon."

Gene felt a prickling sensation on his scalp. The paths of Charles Monet and Peter Cardinal had crossed at only one place on earth, and that was inside Kitum Cave. What had they done in the cave? What had they found in there? What had they touched? What had they breathed? What lived in Kitum Cave?

Going Deep

Eugene Johnson sat at a picnic table at Fort Detrick, near a duck pond, leaning forward and gazing at me. It was a hot day in the middle of summer. He was wearing sunglasses. He placed his large elbows on the table, took off his sunglasses, and rubbed his eyes. He was six foot two, maybe two hundred and fifty pounds. His eyes were brown and set deep in his bearded face, and there were dark circles under the lower lids. He looked tired.

"So Peter Tukei got on the phone to tell me that the boy had visited Kitum Cave," Johnson said. "I still get chills when I think about it. A few weeks later, I flew to Nairobi, and I talked with David Silverstein, the kid's doctor. Peter Tukei was with me. Then we went everyplace in Kenya the kid went, even to his house. His parents had a nice house in Kisumu. Near Lake Victoria. It was a stucco house with a wall around it, and there was a cook and groundskeepers and a driver. The house was clean and neat, open and white-washed. We saw that there was a rock hyrax living on the roof. It was a pet, and it lived in the gutters. There were a couple of storks, and there were rabbits and goats and all kinds of birds. I didn't see any bats around the house."

He paused, thinking. No one else was around. A few ducks swam in the pond. "I was really nervous about talking with the parents," he said. "See, my wife and I don't have

children. I'm not the kind of a guy who can console a
mother, plus I work for the U.S. military. I had no idea how
to talk to them. I tried to put myself in their place, and I
remembered how I felt when my father died. I let them talk
about their son. Peter Cardinal and his sister had been insep-
arable from the moment he arrived in Kenya. The kids had
spent the whole time together, doing everything together. So
what was the difference in behavior? Why did Peter Cardinal
get the virus but not his sister? I learned there was one
difference in their behavior. The parents told me a story
about the rocks in the cave. They told me their kid was an
amateur geologist. There was this issue: did he cut his hand
on any crystals in the cave? We went over that possibility
with the parents. Peter had said to them that he wanted to
collect some of the crystals in Kitum Cave. So he beat on the
walls of the cave with a hammer and collected some rocks
with crystals in them. The rocks were broken up by the
driver and washed by the cook. We tested *their* blood, and
they were not positive for Marburg.''

It seemed possible that the point of contact had been the
boy's hands, that the virus had entered Cardinal's bloodstream through a tiny cut. Possibly he had pricked his finger
on a crystal that had been contaminated with urine from
some animal or the remains of a crushed insect. But even if
he had pricked his finger on a crystal, that didn't tell where
the virus lived in nature; it didn't identify the virus's natural
host.

"We went to look at the cave," he said. "We had to
protect ourselves when we went inside. We knew that Marburg is transmitted by the aerosol route.''

In 1986—the year before Peter Cardinal died—Gene
Johnson had done an experiment that showed that Marburg
and Ebola can indeed travel through the air. He infected
monkeys with Marburg and Ebola by letting them breathe it
into their lungs, and he discovered that a very small dose of
airborne Marburg or Ebola could start an explosive infection

in a monkey. Therefore, Johnson wanted the members of the expedition to wear breathing apparatus inside the cave.

"I brought with me these military gas masks with filters. We needed some kind of covering to put over our heads, too, or we'd get bat shit in our hair. We bought pillowcases at a local store. They were white, with big flowers. So the first time we went into the cave, it was a bunch of Kenyans and me wearing these military gas masks and these flowered pillowcases on our heads, and the Kenyans are just cracking up.''

They explored the cave and made a map of it. After this scouting trip, Gene Johnson persuaded the Army to sponsor a major expedition to Kitum Cave. Half a year after Peter Cardinal died, in the spring of 1988, Gene showed up in Nairobi with twenty shipping crates full of biohazard gear and scientific equipment. It included several military body bags, for holding human cadavers, and the members of his team had a serious discussion among themselves about how to handle their own remains if one of them died of Marburg. This time, Gene felt that he was closing in on the virus. He knew it would be hard to find even if it lived inside Kitum Cave, but he felt he was getting too close to fail in his quest. The monster lived in a cave, and he was going in there to find it.

The Kenyan government agreed to close Kitum Cave to tourists while the joint Kenya-U.S. expedition searched it for viruses. The head of the expedition was Dr. Peter Tukei of the Kenya Medical Research Institute. Gene Johnson conceived the idea and gathered the equipment and found the money to pay for it. There were thirty-five team members, and most of them were Kenyans, including wildlife naturalists, scientists, doctors, and laborers. They brought along a large number of guinea pigs, traveling in boxes, and seventeen monkeys in cages, including baboons, Sykes' monkeys, and African green monkeys. The monkeys and guinea pigs were sentinel animals, like canaries in a coal mine: they

would be placed in cages inside and near Kitum Cave in the
hope that some of them would break with Marburg virus.
There are no instruments that can detect a virus. The best
way to find a virus in the wild, at the present time, is to place
a sentinel animal at the suspected location of the virus and
hope the animal gets sick. Johnson figured that if any of his
monkeys or guinea pigs crashed, he would be able to isolate
the virus from the sick animals and would perhaps be able to
discover how the animals had caught it.

1988 SPRING

The Kitum Cave expedition set up headquarters in the
Mount Elgon Lodge, a decayed resort dating from the nineteen twenties, when the English had ruled East Africa. The
lodge had been built for sporting people and trout fishermen.
It sat on a promontory overlooking the red-dirt road that
wound up the mountain to Kitum Cave. It had once been
surrounded by English gardens, which had partly collapsed
into clay and African weeds. Indoors there were hardwood
floors, waxed daily to a perfect gleam. The lodge had turrets
with round rooms and medieval doors, hand-carved from
African olive wood, and the living room boasted an immense
fireplace with a carved mantelpiece. The staff spoke very
little English, but they were intent on maintaining English
hospitality for the rare guest who might happen to show up.
The Mount Elgon Lodge was a monument to the incomplete
failure of the British Empire, which carried on automati-
cally, like an uncontrollable tic, in the provincial backwaters
of Africa long after it had died at the core. In the evenings,
as the frost-tinged night came on, the staff built fires of
Elgon olive logs in the fireplaces, and the food in the dining
room was horrible, in the best English tradition. There was,
however, a splendid bar. It was a quaint hideaway in a round
chamber, stocked with shining rows of Tusker-beer bottles
and French aperitifs and obscure African brandies. The men

could sit at the bar and drink Tuskers or lean on the great mantel by the fire and tell stories after a hard day in the cave wearing a space suit. A sign on the wall by the concierge's desk mentioned the delicate matter of money. It announced that since the Mount Elgon Lodge's suppliers had cut off all credit to the lodge, the lodge was *unfortunately* unable to extend any credit to its guests.

They moved the animals up the mountain in stages, to let them get used to the climate. When they got to the valley that leads to the cave, they cleared away some underbrush and put up blue tarpaulins. The cave itself was considered to be a Level 4 hot zone. The tarp closest to the cave covered a gray area, a place where the worlds met. The men took chemical showers under the gray-area tarp, to decon their space suits after a visit to the cave. Another tarp covered a Level 3 staging area, where the men changed in and out of their space suits. Another tarp covered a Level 4 necropsy area. Under that tarp, wearing space suits, they dissected any small animals they had trapped, looking for signs of Marburg virus.

"We were going where no one had gone before," Johnson said to me. "We brought the Biosafety Level 4 philosophy to the jungle."

They wore orange Racal space suits inside the cave. A Racal suit is a portable, positive-pressure space suit with a battery-powered air supply. It is for use in fieldwork with extreme biohazards that are believed to be airborne. A Racal suit is also known as an orange suit because it is bright orange. It is lighter than a Chemturion, and unlike a Chemturion, it is fully portable, with a self-contained breathing apparatus. The main body of the suit (apart from the helmet and the blowers) is disposable, so that you can burn it after using it once or twice.

Wearing their Racal space suits, they laid out a trail that wound into Kitum Cave, marking the trail with avalanche poles so that people would not get lost. Along the trail, they placed cages holding the monkeys and guinea pigs. They surrounded the cages with electrified wire, powered by a battery, to discourage leopards from trying to eat the monkeys. They placed some of the monkeys directly underneath bat colonies in the roof of the cave, hoping that something would drop on a monkey that would cause the animal to break with Marburg.

They collected somewhere between thirty thousand and seventy thousand biting insects inside the cave—the cave is full of bugs. "We put stickum paper over cracks in the cave, to catch crawling bugs," Johnson said to me. "We hung light traps inside the cave to collect flying insects. The light traps were battery powered. You know how to collect ticks? They come out of the ground when they smell carbon dioxide from your breath. They smell it and come up and bite your ass. So we brought these huge tanks of carbon dioxide, and we used it to attract ticks. We trapped all the rodents that went into the cave. We used Havahart traps. Way at the back of the cave, by a pool of water, we found sand flies. These are biting flies. We saw leopard tracks all over the place, and Cape-buffalo tracks. We didn't take any blood samples from large animals, nothing from leopards or buffalo. Nothing from the antelopes."

"Could Marburg live in large African cats?" I asked. "Could it be a leopard virus?"

"Maybe. We just didn't have permits to trap leopards. We did collect genet cats, and it wasn't there."

"Could it live in elephants?"

"Did you ever try to draw blood from a wild elephant? We didn't."

The Kenyan naturalists trapped and netted hundreds of birds, rodents, hyraxes, and bats. In the hot necropsy zone, under the tarp, they sacrificed the animals and dissected them while wearing Racal suits, taking samples of blood and tissue, which they froze in jars of liquid nitrogen. Some local people—they were Elgon Masai—had lived inside some of the caves on Mount Elgon and had kept their cattle in the caves. The Kenyan doctors drew blood from these people and took their medical histories, and drew blood from their cattle. None of the local people or the cattle tested positive for Marburg antibodies—if they had tested positive, it would have shown that they had been exposed to Marburg. Despite the fact that nobody showed signs of having been infected, the Elgon Masai could tell stories of how a family member, a child or a young wife, had died bleeding in someone's arms. They had seen family members crash and bleed out, but whether their illnesses were caused by Marburg or some other virus—who could tell? Perhaps the local Masai people knew the Marburg agent in their own way. If so, they had never given it a name.

None of the sentinel monkeys became sick. They remained healthy and bored, having sat in their cages in the cave for weeks. The experiment required that they be sacrificed at the end of the time so that the researchers could take tissue samples and observe their bodies for any signs of infection. At this point, the hard part of primate research began to torment Gene Johnson. He could not bring himself to euthanize the monkeys. He couldn't stand the idea of killing them and couldn't go into the cave to finish the job. He waited outside in the forest while another member of the team put on a space suit and went inside and gave the monkeys massive shots of sedative, which put them to sleep forever. "I don't like killing animals," he said to me. "That was a major issue for me. After you've fed and watered monkeys for thirty days, they become your friends. I fed 'em bananas. That was terrible. It sucked." He put on his orange Racal space suit and opened up the monkeys under the necropsy tent, feeling frustrated and sad, especially when all the monkeys turned out to be healthy.

The expedition was a dry hole. All of the sentinel animals remained healthy, and the blood and tissue samples from the other animals, insects, birds, Masai people, and their cattle showed no signs of Marburg virus. It must have been a bitter disappointment for Gene Johnson, so disheartening that he was never able to bring himself to publish an account of the expedition and its findings. There seemed to be no point in publishing the fact that he hadn't found anything in Kitum Cave. All that he could say for sure is that Marburg lives in the shadow of Mount Elgon.

What Johnson did not know at the time, but what he sensed almost instinctively after the failure of the Kitum Cave expedition, was that the knowledge and experience he gained inside a cave in Africa, and the space suits and bio-hazard gear he carried back with him to Fort Detrick, might serve him well at another time and in another place. He kept his African gear hidden away at the Institute, piled in olive-drab military trunks in storage rooms and in tractor trailers parked behind buildings and padlocked, because he did not want anyone else to touch his gear or use it or take it away from him. He wanted to be ready to use it at a moment's notice, in case Marburg or Ebola ever came to the surface again. And sometimes he thought of a favorite saying, a remark by Louis Pasteur, ''Chance favors the prepared mind.'' Pasteur developed vaccines for anthrax and rabies.

1989 SUMMER

The Army had always had a hard time figuring out what to do with Nancy and Jerry Jaax. They were married officers at the same rank in a small corps, the Veterinary Corps. What if one of them (the wife) is trained in the use of space suits? Where do you send them? The Army assigned the Jaaxes to the Institute of Chemical Defense, near Aberdeen, Maryland. They sold their Victorian house and moved, bringing their birds and animals with them. Nancy was not sorry to leave the house in Thurmont. They moved into a tract house, which was more to her liking, and there they began to raise fish in tanks, as a hobby, and Nancy went to work in an

Army program to study the effects of nerve gas on rat brains. Her job was to open up the rat's head and figure out what the nerve gas had done to the brain. This was safer and more pleasant than working with Ebola, but it was a little dull. Eventually she and Jerry both received promotions to lieutenant colonel and wore silver oak leaves on their shoulders. Jaime and Jason were growing up. Jaime became a superb gymnast, short and wiry like Nancy, and Nancy and Jerry had hopes for her in the nationals, if not the Olympics. Jason grew into a tall, quiet kid. Herky, the parrot, did not change. Parrots live for many years. He went on shouting "Mom! Mom!" and whistling the march from *The Bridge on the River Kwai.*

Colonel Tony Johnson, Nancy's commanding officer when she had worked at USAMRIID, remembered her competence in a space suit and wanted to get her back. He felt she belonged at the Institute. He was eventually appointed head of pathology at Walter Reed Army Medical Center, and when that happened, his old job came open, the job of chief of pathology at the Institute. He urged the Army to appoint Nancy Jaax to the position, and the Army listened. They agreed that she ought to be doing hot biological work, and she got the job in the summer of 1989. At the same time, the Army appointed Jerry Jaax head of the veterinary division at the Institute. So the Jaaxes became important and rather powerful figures. Nancy went back to biological work in space suits. Jerry still didn't like it, but he had learned to live with it.

With these promotions, the Jaaxes sold their house in Aberdeen and moved back to Thurmont, in August 1989. This time, Nancy told Jerry it was not going to be a Victorian. They bought a contemporary Cape house with dormer windows, with a lot of land around it, meadow and forest, where the dogs could run and the children could play. Their house stood on the lower slope of Catoctin Mountain, overlooking the town, above a sea of apple orchards. From their kitchen

window, they could look into the distance over rolling farmland where armies had marched during the Civil War. Central Maryland stretched away to the horizon in folds and
hollows, in bands of trees and rumpled fields, studded by
silos that marked the presence of family farms. High over the
beautiful countryside, passenger jets crisscrossed the sky,
leaving white contrails behind them.

PART TWO

THE
MONKEY HOUSE

Reston

The city of Reston, Virginia, is a prosperous community about ten miles west of Washington, D.C., just beyond the Beltway. On a fall day, when a western wind clears the air, from the upper floors of the office buildings in Reston you can see the creamy spike of the Washington Monument, sitting in the middle of the Mall, and beyond it the Capitol dome. Reston was one of the first planned suburbs in America, a visible symbol of the American belief in rational design and suburban prosperity, a community of gently curved streets, making arcs through landscaped neighborhoods, where disorder and chaos were given no sign of acknowledgment and no places to hide. The population of Reston has grown in recent years, and high-technology businesses and blue-chip consulting firms have moved into office parks there, where glass buildings grew up during the nineteen-eighties like crystals. Before the crystals appeared, Reston was surrounded by farmland, and the town still contains meadows. In spring, the meadows burst into galaxies of yellow-mustard flowers, and robins and thrashers sing in stands of tulip trees and white ash. The town offers handsome, expensive residential neighborhoods, good schools, parks, golf courses, excellent day care for children. There are lakes in Reston named for American naturalists (Lake Thoreau, Lake Audubon), surrounded by water-front homes. Reston is situated within easy commuting distance of downtown Washington. Along Leesburg Pike, which funnels traffic into the city, there are developments of executive homes with Mercedes-Benzes parked in crescent-shaped driveways. Reston was once a country town, and its rural past still fights obliteration, like a nail that won't stay hammered down. Among the upscale houses, you see the occasional bungalow with cardboard stuffed in a broken window and a pickup truck parked in the side yard. In the autumn, vegetable stands along Leesburg Pike sell pumpkins and butternut squash.

Not far from Leesburg Pike there is a small office park. It was built in the nineteen-sixties, and is not as glassy or as fashionable as the newer office parks, but it is clean and neat, and it has been there long enough for sycamores and sweet-gum trees to grow up around it and throw shade over the lawns. Across the street, a McDonald's is jammed at lunch hour with office workers. In the autumn of 1989, a company called Hazleton Research Products was using a one-story building in the office park as a monkey house. Hazleton Research Products is a division of Corning, Inc. Corning's Hazleton unit is involved with the importation and sale of laboratory animals. The Hazleton monkey house was known as the Reston Primate Quarantine Unit.

Each year, about sixteen thousand wild monkeys are imported into the United States from the tropical regions of the earth. Imported monkeys must be held in quarantine for a month before they are shipped anywhere else in the United States. This is to prevent the spread of infectious diseases that could kill other primates, including humans.

Dan Dalgard, a doctor of veterinary medicine, was the consulting veterinarian at the Reston Primate Quarantine Unit. He was on call to take care of the monkeys if they became sick or needed medical attention. He was actually a principal scientist at another company owned by Corning, called Hazleton Washington. This company has its head-quarters on Leesburg Pike in Vienna, Virginia, not far from the monkey house in Reston, and so Dalgard could easily drive his car over to Reston to check on the monkeys if he was needed there. Dalgard was a tall man in his fifties, with metal-framed glasses, pale blue eyes, a shy manner, and a soft drawl that he had picked up in Texas at veterinary school. Generally he wore a gray business suit if he was working in his office, or a white lab coat if he was working with animals. He had an international reputation as a knowl-edgeable and skilled veterinarian who specialized in primate husbandry. He was a calm, even-tempered man with a kind of dreamy nature; a man given to staring out the window of his office, thinking about one thing or another. On evenings and weekends, he repaired antique clocks as a hobby. He liked to fix things with his hands; it made him feel peaceful and calm and daydreamy, and he was patient with a jammed clock. He sometimes had longings to leave veterinary medi-cine and devote himself full-time to clocks.

On Wednesday, October 4, 1989, Hazleton Research Products accepted a shipment of a hundred wild monkeys from the Philippines. The shipment originated at Ferlite Farms, a monkey wholesale facility located not far from the city of Manila. The monkeys themselves came from coastal rain forests on the island of Mindanao. The monkeys had been shipped by boat to Ferlite Farms, where they were jammed together in large cages known as gang cages, in which the male monkeys often fought and bloodied and killed one another. The monkeys were then put into wooden crates and flown to Amsterdam on a specially fitted cargo airplane, and from Amsterdam they were flown to New York City. They arrived at JFK International Airport and were driven by truck down the eastern seaboard of the United States to the Reston monkey house.

The monkeys were crab-eating monkeys, a species that

lives along rivers and in mangrove swamps in Southeast Asia. Crab eaters are used as laboratory animals because they are common, cheap, and easily obtained. They have long, arching, whiplike tails, whitish fur on the chest, and cream-colored fur on the back. The crab eater is a type of macaque (pronounced ma-KACK). It is sometimes called a long-tailed macaque. The monkey has a protrusive, doglike snout with flaring nostrils and exceedingly sharp canine teeth, able to rip flesh as easily as a honed knife. The skin is pinkish gray, close to the color of a white person. The hand looks quite human, with a thumb and delicate fingers with fingernails. The females have two breasts on the upper chest that look startlingly human, with pale nipples.

Crab eaters do not like humans. They have a competitive relationship with people who live in the rain forest. They like vegetables, especially eggplants, and they like to raid farmers' crops. Crab-eating monkeys travel in a troop, making tumbling jumps through the trees, screaming, *"Kra! Kra!"* They know perfectly well that after they have pulled off an eggplant raid on a farmer's field they are likely to have a visit from the farmer, who will come around looking for them with a shotgun, and so they have to be ready to move out and head deep into the forest at a moment's notice. The sight of a gun will set off their alarm cries: *"Kra! Kra! Kra!"* In some parts of the world, these monkeys are called *kras,* because of the sound they make, and many people who live in Asian rain forests consider them to be obnoxious pests. At the close of day, when night comes, the troop goes to sleep in a dead, leafless tree. This is the troop's home tree. The monkeys prefer to sleep in a dead tree so that they can see in all directions, keeping watch for humans and other evil predators. The monkey tree usually hangs out over a river, so that they can relieve themselves from the branches without littering the ground.

At sunrise, the monkeys stir and wake up, and you hear

their cries as they greet the sun. The mothers gather their children and herd them along the branches, and the troop moves out, leaping through trees, searching for fruit. They like to eat all kinds of things. In addition to vegetables and fruits, they eat insects, grass, roots, and small pieces of clay, which they chew and swallow, perhaps to get salt and minerals. They lust after crabs. When the urge for crabs comes upon them, the troop will head for a mangrove swamp to have a feeding bout. They descend from the trees and take up positions in the water beside crab holes. A crab comes out of its hole, and the monkey snatches it out of the water. The monkey has a way to deal with the crab's claws. He grabs the crab from behind as it emerges from its hole and rips off the claws and throws them away and then devours the rest of the crab. Sometimes a monkey isn't quick enough with the claws, and the crab latches onto the monkey's fingers, and the monkey lets out a shriek and shakes its hand, trying to get the crab off, and jumps around in the water. You can always tell when crab eaters are having a feeding bout on crabs because you hear an occasional string of shrieks coming out of the swamp as a result of difficulty with a crab.

The troop has a strict hierarchy. It is led by a dominant male, the largest, most aggressive monkey. He maintains control over the troop by staring. He stares down subordinates if they challenge him. If a human stares at a dominant male monkey in a cage, the monkey will rush to the front of the cage, staring back, and will become exceedingly angry, slamming against the bars, trying to attack the person. He will want to kill the human who stared at him: he can't afford to show fear when his authority is challenged by another evil primate. If two dominant male monkeys are placed in the same cage, only one monkey will leave the cage alive.

The crab-eating monkeys at the Reston monkey house were placed each in its own cage, under artificial lights, and were fed monkey biscuits and fruit. There were twelve monkey rooms in the monkey house, and they were designated by the letters A through L. Two of the monkeys that arrived on October 4 were dead in their crates. That was not unusual, since monkeys die during shipments. But in the next three weeks, an unusual number of monkeys began to die at the Reston monkey house.

On October 4, the same day the shipment of monkeys reached the Reston monkey house, something happened that would change Colonel Jerry Jaax's life forever. Jerry had a brother named John, who lived in Kansas City with his wife and two small children. John Jaax was a prominent business-man and a banker, and he was a partner in a manufacturing company that made plastic for credit cards. He was a couple of years younger than Jerry, and the two men were as close as brothers can be. They had grown up together on a farm in Kansas and had both gone to college at Kansas State. They looked very much alike: tall, with prematurely gray hair, a beak nose, sharp eyes, a calm manner; and their voices sounded alike. The only difference in appearance between them was that John wore a mustache and Jerry did not.

John Jaax and his wife planned to attend a parent-teach-ers' meeting on the evening of October 4 at their children's school. Near the end of the day, John telephoned his wife from his office at the manufacturing plant to tell her that he would be working late. She happened to be out of the house when he called, so he left a message on the answering ma-chine, explaining that he would go directly from the office to the meeting, and he would see her there. When he did not show up, she became worried. She drove over to the factory.

The place was deserted, the machines silent. She walked the length of the factory floor to a staircase. John's office overlooked the factory floor from a balcony at the top of the staircase. She climbed the stairs. The door to his office was standing open a crack, and she went inside. John had been shot many times, and there was blood all over the room. It was a violent killing.

The police officer who took the case at Kansas City Homicide was named Reed Buente. He had known John personally and had admired him, having worked for him as a security guard at the Bank of Kansas City when John was president of the bank. Officer Buente was determined to solve the case and bring the killer or killers to trial. But as time went by and no breaks came along, the investigator became discouraged. John Jaax had been having difficulties with his partner in the plastic business, a man named John Weaver, and Kansas City Homicide looked at the partner as a suspect. (When I called Officer Buente recently, he confirmed this. Weaver has since died of a heart attack, and the case remains open, since unsolved murder cases are never closed.) There were few physical clues, and Weaver, as it turned out, had an alibi. The investigator ran into more and more difficulties with the case. At one point, he said to Jerry, "You can have someone killed pretty easy. And it's cheap. You can have someone killed for what you would pay for a desk."

The murder of John Jaax threw Jerry into a paralysis of grief. Time is supposed to heal all things, but time opened an emotional gangrene in Jerry. Nancy began to think that he was in a clinical depression.

"I feel like my life is over," he said to her. "It's just not the same anymore. My life will never be the same. It's just inconceivable that Johnny could have had an enemy." At the funeral in Kansas City, Nancy and Jerry's children, Jaime and Jason, looked into the coffin and said to their father, "Gee, Dad, he looks like you lying there."

Jerry Jaax called Kansas City Homicide nearly every day during October and November. The investigator just couldn't break the case. He began to think about getting a gun and going out to Kansas City to kill John's business partner. He thought, If I do it, I'll be in jail, and what about my children? And what if John's partner hadn't been behind the murder? Then I'll have killed an innocent man.

NOVEMBER 1, WEDNESDAY

The colony manager at the Reston monkey house will be called Bill Volt. As he watched his monkeys die, Volt became alarmed. On November 1, a little less than a month after the shipment of monkeys had arrived, he put in a telephone call to Dan Dalgard, telling him that the monkeys that had recently arrived from the Philippines were dying in unusually large numbers. He had counted twenty-nine deaths out of a shipment of a hundred monkeys. That is, nearly a third of the monkeys had died. At the same time, a problem had developed with the building's heating and air-handling system. The thermostat had failed, and the heat would not go off. The heaters dumped heat at full blast into the building, and the air-conditioning system would not kick in. It had become awfully hot inside the building. Volt wondered if the heat might be putting stress on the monkeys. He had noticed that most of the deaths had taken place in one room, Room F, which was located on a long hallway at the back of the building.

Dalgard agreed to drive over to the monkey house and have a look, but he became busy with other things and did not get there until the following week. When he arrived, Bill Volt took him to Room F, the focus of the deaths, so that Dalgard could inspect the monkeys. They put on white coats and surgical masks, and the two men walked down a long cinder-block corridor lined on both sides with steel doors leading to monkey rooms. The corridor was very warm, and they began to sweat. Through windows in the doors, they could see hundreds of monkey eyes looking at them as they passed. The monkeys were exquisitely sensitive to the presence of humans.

Room F contained only crab-eating monkeys from the October shipment from Ferlite Farms in the Philippines. Each monkey sat in its own cage. The monkeys were subdued. A few weeks ago, they had been swinging in the trees, and they didn't like what had happened to them. Dalgard went from cage to cage, glancing at the animals. He could tell a lot about a monkey from the look in its eyes. He could also read its body language. He searched for animals that seemed passive or in pain.

Dalgard's staring into their eyes drove them berserk. When he passed a dominant male and looked carefully at it, it rushed him, wanting to take him out. He found a monkey whose eyes had a dull appearance, not shiny and bright but glazed and somewhat inactive. The eyelids were down, slightly squinted. Normally the lids would be retracted so that you could see the entire iris. A healthy monkey's eyes would be like two bright circles in the monkey's face. This animal's eyelids had closed down slightly, and they drooped, so that the iris had become a squinting oval. That was a sign of illness in the monkey.

He put on leather gauntlet gloves, opened the door of the cage, reached inside, and pinned the monkey down. He slipped one hand out of a glove and quickly felt the monkey's stomach. Yes—the animal felt warm to the touch. So it had a fever. And it had a runny nose. He let go of the monkey and shut the door. He didn't think that the animal was suffering from pneumonia or a cold. Perhaps the animal was affected by heat stress. It was very warm in this room. He advised Bill Volt to put some pressure on the landlord to get the heating system fixed. He found a second animal that also had droopy eyelids, with that certain squint in the eyes. This one also felt hot to the touch, feverish. So there were two sick monkeys in Room F.

Both monkeys died during the night. Bill Volt found them in the morning, hunched up in their cages, staring with glassy, half-open eyes. This greatly concerned Volt, and he decided to dissect the animals, to try to see what had killed them. He carried the two deceased monkeys into an examination room down the hallway and shut the door after him, out of sight of the other monkeys. (You can't cut up a dead monkey in front of other monkeys—it will cause a riot.) He opened the monkeys with a scalpel and began his inspection. He did not like what he saw, and did not understand it, so he called Dalgard on the telephone and said, "I wonder if you could come over here and have a look at these monkeys."

Dalgard drove over to the monkey house immediately. His hands, which were so confident and skillful at taking apart clocks, probed the monkeys. What he saw inside the animals puzzled him. They appeared to have died of heat stress, brought on, he suspected, by the problems with the heating system in the building—but their spleens were weirdly enlarged. Heat stress wouldn't enlarge the spleen, would it? He noticed something else that gave him pause. Both animals had small amounts of blood in their intestines. What could do that?

Later that same day, another large shipment of crab-eating monkeys arrived from Ferlite Farms. Bill Volt put the new monkeys in Room H, two doors down the hall from Room F.

Dan Dalgard became very worried about the monkeys in Room F. He wondered if there was some kind of infectious agent going around the room. The blood in the gut looked like the effects of a monkey virus called simian hemorrhagic fever, or SHF. This virus is deadly to monkeys, although it is harmless to people. (It can't live in humans.) Simian fever can spread rapidly through a monkey colony and will generally wipe it out.

It was now Friday, November 10. Dalgard planned to spend the weekend fixing his clocks in the family room of his house. But that Saturday morning at home, as he laid out his tools and the pieces of an antique clock that needed fixing, he could not stop thinking about the monkeys. He was worried about them. Finally he told his wife that he had to go out on company business, and he put on his coat and drove over to the monkey house and parked in front of the building and went in through the front door. It was a glass door, and as he opened it, he felt the unnatural heat in the building wash over him, and he heard the familiar screeches of monkeys. He went into Room F. *"Kra! Kra!"* the monkeys cried at him in alarm. There he found three more dead monkeys. They were curled up in their cages, their eyes open, expressionless. This was not good. He carried the dead monkeys into the examination room and slit the animals open, and looked inside.

Soon afterward, Dan Dalgard began to keep a diary. He kept it on a personal computer, and he would type in a few words each day. Working quickly and without much thought, he gave his diary a title, calling it, "Chronology of Events." It was now getting close to the middle of November, and as the sun went down in the afternoon and traffic jams built up on Leesburg Pike near his office, Dalgard worked on his diary. Tapping at the keys, he would later recall in his mind's eye what he had seen inside the monkeys.

> The lesions by this time were showing a pattern of marked splenomegaly [swollen spleen]—strikingly dry on cut surface, enlarged kidneys, and sporadic occurrences of hemorrhage in a variety of organs. . . . Clinically, the animals showed abrupt anorexia [loss of appetite], and lethargy. When an animal began showing signs of anorexia, its condition deteriorated rapidly. Rectal temperatures taken on monkeys being sacrificed were not elevated. Nasal discharge, epistaxis [bloody nose] or bloody stools were not evident. . . . Many of the animals were in prime condition and had more body fat than is customary for animals arriving from the wild.

There was nothing much wrong with the dead animals, nothing that he could put his finger on. They simply stopped eating and died. They died with their eyes open, and with staring expressions on their faces. Whatever this disease was, the cause of death was not obvious. Was it a heart attack? A fever? What?

The spleen was inexplicably damaged. The spleen is a kind of bag that filters the blood, and it plays a role in the immune system. A normal spleen is a soft sack with a drippy red center, which reminded Dalgard of a jelly doughnut. When you cut into a normal spleen with a scalpel, it gives about as much resistance to the knife as a jelly doughnut, and it drips a lot of blood. But these spleens had swelled up and turned as hard as a rock. A normal monkey spleen would be about the size of a walnut. These spleens were the size of a tangerine and were leathery. They reminded him of a piece of salami—meaty, tough, dry. His scalpel practically bounced off them. He could actually tap the blade of the scalpel on the spleen, and the blade wouldn't dig in very much. What he didn't realize—what he couldn't see because it was almost inconceivable—was that the entire spleen had become a solid clot of blood. He was tapping his scalpel on a blood clot the size of a tangerine.

On Sunday, November 12, Dalgard puttered around the house in the morning, fixing things, doing little errands. After lunch, he once again returned to the monkey house. He found three more dead monkeys in Room F. They were dying steadily, a handful every night. There was a mystery developing in the Reston Primate Quarantine Unit.

One of the dead animals had been given the name O53. Dalgard carried the carcass of Monkey O53 into the examination room and opened it up and looked inside the body cavity. With a scalpel, he removed a piece of Monkey O53's spleen. It was huge, hard, and dry. He took a Q-Tip and rubbed it in the dead monkey's throat, collecting a little bit of mucus, a throat wash. Then he swirled the Q-Tip in a test tube full of distilled water and capped the tube. Anything alive in the mucus would be preserved temporarily.

Into Level 3

By Monday morning—the day after he dissected Monkey O53—Dan Dalgard had decided to bring the problem with his monkeys to the attention of USAMRIID, at Fort Detrick. He had heard that the place had experts who could identify monkey diseases, and he wanted to get a positive identification of the sickness. Fort Detrick was about an hour's drive northwest of Reston, on the other side of the Potomac River.

Dalgard ended up talking by phone with a civilian virologist named Peter Jahrling. Jahrling had a reputation for knowing something about monkey viruses. They had never talked before. Dalgard said to Jahrling, "I think we've got some SHF [simian hemorrhagic fever] in our monkeys. The spleen looks like a piece of salami when you slice it." Dalgard asked Jahrling if he would look at some samples and give a diagnosis, and Jahrling agreed to help. The problem attracted Peter Jahrling's curiosity.

Jahrling had worked at the Institute for most of his career, after an early period in which he had lived in Central America and hunted for viruses in the rain forest (he had discovered several previously unknown strains). He had blond hair, beginning to go gray, steel-rimmed glasses, a pleasant, mobile face, and a dry sense of humor. He was by nature a

cautious, careful person. Peter Jahrling spent large amounts
of time wearing a Chemturion biological space suit. He performed research on defenses against hot viruses—vaccines,
drug treatments—and he did basic medical research on rain-
forest viruses. The killers and the unknowns were his spe-
cialty. He deliberately kept his mind off the effects of hot
agents. He told himself, If you did think about it, you might
decide to make a living another way.

Jahrling, his wife, and their three children lived in Thurmont, not far from Nancy and Jerry Jaax, in a brick ranch
house with a white picket fence out front. The fence surrounded a treeless yard, and there was a large brown car
parked in the garage. Although they lived near each other,
the Jahrlings did not socialize with the Jaaxes, since their
children were of different ages and since the families had
different styles.

Peter Jahrling mowed his lawn regularly to keep the grass
neat, so that his neighbors wouldn't think he was a slob.
Externally he lived a nearly featureless life among suburban
neighbors, and very few of them knew that when he climbed
into his mud-colored car he was headed for work in a hot
zone, although the license plate on the car was a vanity plate
that said LASSA. Lassa is a Level 4 virus from West Africa,
and it was one of Peter Jahrling's favorite life forms—he
thought it was fascinating and beautiful, in certain ways. He
had held in his gloved hands virtually every hot agent
known, except for Ebola and Marburg. When people asked
him why he didn't work with those viruses, he replied, "I
don't particularly feel like dying."

After his telephone conversation with Dan Dalgard, Peter
Jahrling was surprised and annoyed when, the next day, a
few bits of frozen meat from Monkey O53 arrived at the
Institute, brought by courier. What annoyed him was the fact
that the bits of meat were wrapped in aluminum foil, like
pieces of leftover hot dog.

The hot-dog-like meat was monkey spleen, and the ice around it was tinged with red and had begun to melt and drip. The samples also included the tube containing the throat wash and some blood serum from the monkey. Jahrling carried the samples into a Level 3 laboratory. Level 3 is kept under negative air pressure, to prevent things from leaking out, but you don't need to wear a space suit there. People who work in Level 3 dress themselves like surgeons in an operating room. Jahrling wore a paper surgical mask, a surgical scrub suit, and rubber gloves. He peeled off the tin foil. A pathologist helped him do it, standing next to him. The bit of spleen rolled about on the tin foil as they poked it—a hard little pink piece of meat, just as Dalgard had described it. Jahrling thought, Like the kind of mystery meat you get in a school lunchroom. Jahrling turned to the other man and remarked, "Good thing this ain't Marburg," and they chuckled.

Later that day, he called Dalgard on the telephone and said to him something like, "Let me tell you how to send a sample to us. People around here may be slightly paranoid, but they get a little upset when you send a sample and it drips blood on the carpet."

One way to identify a virus is to make it grow inside living cells in a flask of water. You drop a sample of the virus into the flask, and the virus spreads through the cells. If the virus likes the cells, it will multiply. One or two viruses can become a billion viruses in a few days—a China of viruses in a bottle the size of one's thumb.

A civilian technician named Joan Rhoderick cultured the unknown agent from Monkey O53. She ground up a bit of the monkey's spleen with a mortar and pestle. That made sort of a bloody mush. She dropped the mush into flasks that contained living cells from the kidney of a monkey. She also took some of the throat mucus from Monkey O53 and put it into a flask, and she took some of the monkey's blood serum and put it into another flask. Eventually she had a whole rack of flasks. She put them into a warmer—an incubator, held at body temperature—and hoped that something would grow. Growing up a virus in culture is a lot like making beer. You follow the recipe, and you keep the brew nice and warm until something happens.

Dan Dalgard did not visit the monkey house the next day, but he telephoned Bill Volt, the manager, to find out how things were going. Volt reported that all the animals looked good. None of them had died during the night. The illness seemed to be fading away naturally. Fortunately, it looked like things were quieting down in Reston, and Dalgard felt relieved that his company had dodged a bullet.

But what were those Army people doing with the samples of monkey? He called Jahrling and learned that it was too soon to know anything. It takes several days to grow up a virus.

A day later, Bill Volt called Dalgard with bad news. Eight monkeys in Room F had stopped eating. In other words, eight monkeys were getting ready to die. The thing had come back.

Dalgard hurried over to the monkey house, where he found that the situation had deteriorated suddenly. There were many more animals with squinting, glazed, oval-shaped eyes. Whatever the thing was, it was steadily working its way through Room F. By now, fully half the animals in the room had died. It was going to kill the entire room if nothing was done to stop it. Dalgard became extremely anxious for some news from Peter Jahrling.

Thursday, November 16, arrived, and with it came news that monkeys had begun to die in rooms down the hallway from Room F. Late in the morning, Dan Dalgard received a telephone call from Peter Jahrling. A pathologist at the Institute had inspected the meat very carefully and had given it a tentative diagnosis of simian hemorrhagic fever—harmless to humans, lethal to monkeys.

Dalgard now knew that he had to move fast to contain the outbreak before the virus spread through the monkey house. Simian hemorrhagic fever is highly contagious in monkeys. That afternoon, he drove up Leesburg Pike to the office park in Reston. At five o'clock on a gray, rainy evening on the edge of winter, as commuters streamed home from Washington, he and another Hazleton veterinarian injected all the monkeys in Room F with lethal doses of anesthetic. It was all over quickly. The monkeys died in minutes.

Dalgard opened up eight healthy-looking carcasses to see if he could find any signs of simian fever inside them. He was surprised to see that there didn't seem to be anything wrong with them. This greatly troubled him. Sacrificing the monkeys had been a difficult, disgusting, and disheartening task. He knew there was a disease in this room, and yet these monkeys were beautiful, healthy animals, and he had just killed them. The sickness had been entrenched in the building since early October, and it was now the middle of November. The Army had given him a tentative diagnosis, probably the best diagnosis he would ever get, and he had been left with the unpleasant task of trying to salvage the lives of the remaining animals. He went home that evening feeling that he had had a very bad day. Later he would write in his diary:

> There was a notable absence of any hemorrhagic component. In general, the animals were unusually well fleshed (butterballs), young (less than 5 years), and in prime condition.

Before he left the monkey house, he and the other veterinarian placed the dead monkeys in clear plastic bags and carried some of them across the hall to a chest freezer. A freezer can be as hot as hell. When a place is biologically hot, no sensors, no alarms, no instruments can tell the story.

All instruments are silent and register nothing. The monkeys' bodies were visible in the clear bags. They froze into contorted shapes, with their chest cavities spread wide and their intestines hanging out and dripping red icicles. Their hands were clenched into fists or open like claws, as if they were grasping at something, and their faces were expressionless masks, their eyes glazed with frost, staring at nothing.

Exposure

NOVEMBER 17, FRIDAY

Thomas Geisbert was an intern at the Institute, a kind of trainee. He was twenty-seven years old, a tall man with dark blue eyes and longish brown hair parted in the middle and hanging over his forehead. Geisbert was a skilled fisherman and a crack shot with a rifle, and he spent a lot of time in the woods. He wore blue jeans and cowboy boots, and tended to ignore authority. He was a local boy who had grown up near Fort Detrick. His father was the chief building engineer at the Institute, the man who repaired and operated the hot zones. When Tom Geisbert was a boy, his father had taken him to visit the Institute, and Tom had stared through the heavy glass windows at people in space suits, thinking it would be cool to do that. Now he was doing it, and it made him happy.

The Institute hired him to operate its electron microscope, which uses a beam of electrons to make images of small objects, such as viruses. It is an essential tool to have around a virus lab because you can use it to make a photograph of a tiny piece of meat and find viruses in the meat. For Geisbert, identifying hot strains and classifying the tribes of viruses was like sorting butterflies or collecting flowers. He liked the loneliness of inner space, the sense of being forgotten by the world. He felt quiet and at peace with himself when he was padding around a hot zone in a space suit, carrying a rack of test tubes that held an unknown deadly agent. He liked to go into the Level 4 suites alone, rather than with a buddy, especially in the middle of the night, but his tendency to spend large amounts of time at his work had begun to affect his personal life, and his marriage was breaking up. He and his wife had separated in September. His troubles at home only reinforced his tendency to bury himself in Level 4.

One of Geisbert's greatest happinesses in life, apart from his work, came from being in the outdoors, fishing for black bass and hunting for deer. He hunted for meat—he gave the venison to members of his family—and then, when he had got the meat he needed, he hunted for trophies. Every year around Thanksgiving, he went hunting in West Virginia, where he and some buddies rented a house for the opening of the deer season. His friends did not know much about what he did for a living, and he made no effort to tell them about it.

Geisbert tried to look at many samples of virus as a way of sharpening his skills with an electron microscope. He was learning how to identify hot agents by eye, by looking at photographs of the particles. When the samples of the Cardinal boy had arrived from Africa, Geisbert had spent days gazing at them. They attracted him. The Cardinal strain was a tangled mass of 6s, Us, gs, Ys, snakes, and Cheerios mixed up with partly liquefied human flesh. Geisbert spent so much time staring at the virus, one of the true horrors of nature, that the shapes became implanted in his mind.

Tom Geisbert had heard about the sick monkeys in Virginia, and he wanted to take photographs of the meat to see if he could identify any simian-fever-virus particles in it. On Friday morning, November 17, the day after Dan Dalgard had killed all the animals in Room F, Geisbert decided to take a look at the flasks of monkey cells that were ripening. He wanted to examine them with a light microscope before he went on his Thanksgiving hunt, to see if he could observe any changes. A light microscope is a standard microscope that uses lenses to focus light.

At nine o'clock on that Friday morning, he put on a surgical scrub suit and a paper mask and went into the Level 3 lab where the flasks were being kept warm. There he met Joan Rhoderick, the technician who had started the Reston culture. She was staring through the binocular eyepieces of the microscope at a small flask. The flask contained cells that had been infected with the simian-fever virus that came from Monkey O53.

She turned to Geisbert. "There's something flaky going on in this flask," she said.

The flask was a typical virus flask. It was about the size of a person's thumb and was made of clear plastic so that you could place it in a microscope and look into the flask. It had a black screw cap.

Geisbert stared through the eyepieces of the microscope. He saw a complicated world in the flask. As always in biology, the problem was to know what you were looking at. The patterns of nature are deep and complex, constantly changing. He saw cells all over the place. They were tiny bags, each containing a nucleus, which was a darker blob near the center. The cells looked a little bit like fried eggs, sunny-side up. The egg yolk would be the cell's nucleus.

Living cells ordinarily stick to the bottom of a flask to form a living carpet—cells prefer to cling to something when they grow. This carpet had been eaten by moths. The cells had died and drifted away, leaving holes in the carpet.

Geisbert checked all the flasks, and most of them looked the same way, like moth-eaten carpet. They looked real bad, they looked sick. Something was killing these cells. They were swollen and puffy, fat looking, as if they were pregnant. Tom could see that they contained granules or specks. The specks looked like pepper. As if someone had shaken pepper over fried eggs. He may have seen reflections of light in the pepper, as if light was gleaming through crystals. Crystals? These cells were unrecognizably sick. And they were very sick, because the fluid was milky and clouded with dead cells, cells that had exploded.

They decided that their boss, Peter Jahrling, should have a look. Geisbert went to find Jahrling. He exited from Level 3—removed his scrub suit and took a water shower and dressed in civilian clothes—and went to Jahrling's office. Then he and Jahrling returned to the Level 3 lab. It took a few minutes for both of them to change in the locker room and put on scrub suits. When they were ready to go in— dressed like surgeons—they entered and sat down at the eye- pieces of the microscope. Geisbert said to him, ''There's something very strange going on in that flask, but I'm not sure what it is. This isn't like SHF.''

Jahrling looked. He saw that the flask had turned milky, as if it had gone rotten. ''This is contaminated,'' he said. ''These cells are blown away. They're crud.'' The cells were exploded and dead. ''They're off the plastic,'' he remarked. By *off the plastic* he meant that the dead cells had detached from the surface of the flask and had floated away in the broth. He thought that a wild strain of bacteria had invaded the cell culture. This is an annoying and common occurrence when you are trying to grow a virus, and it wipes out the flask. The wild bacteria consume the cell culture, eat it up, and make a variety of different smells in the air while they're growing, whereas viruses kill cells without releasing an odor. Jahrling guessed that the flask had been wiped out by a common soil bacterium called pseudomonas. It lives in dirt. It lives in everyone's backyard and under fingernails. It is one of the most common forms of life on the planet, and it often gets into cell cultures and wrecks them.

Jahrling unscrewed the little black cap and waved his hand over the flask to bring the scent to his nose, and then he took a whiff. Hm. Funny. No smell.

He said to Tom Geisbert, "Have you ever smelled pseudomonas?"

"No," Tom replied.

"It smells like Welch's grape juice. Here—" He offered the flask to Tom.

Tom sniffed it. There was no smell.

Jahrling took back the flask and whiffed it again. His nose registered nothing. But the flask was milky, and the cells were blown away. He was puzzled. He handed the flask back to Tom and said, "Put it in the beam, and let's look at it." By *put it in the beam,* he meant "look at it using the electron microscope," which is much more powerful than a light microscope, and can see deeper into the universe within.

Geisbert poured some of the milky fluid out of the flask into a test tube and then spun it in a centrifuge machine. A button of grayish ooze collected at the bottom of the test tube—a tiny pill of dead and dying cells. The pill was the size of a pinhead, and it had a pale brownish color. Geisbert thought it looked like a dab of mashed potato. He lifted out the button with a wooden stick and soaked the button in plastic resin to preserve it. But now, what was on his mind was the hunting season. Later that afternoon—Friday—he went home to get packed. He had been planning to drive his Ford Bronco, but it had broken down; so one of his hunting buddies met him in a pickup truck and they loaded Geisbert's duffel bag and gun case into it and set off on his hunting trip. When a filovirus begins to amplify itself in a human being, the incubation period is from three to eighteen days, while the number of virus particles climbs steadily in the bloodstream. Then comes the headache.

Thanksgiving

For Nancy and Jerry Jaax, it was the worst Thanksgiving of their lives. On Wednesday, November 22, they put their children in the family van and drove straight through the night to Kansas. Jaime was now twelve, and Jason was thirteen. The children were used to long drives to Kansas, and they slept peacefully. Jerry had almost lost his ability to sleep since the murder of his brother, and Nancy stayed awake with him, trading places behind the steering wheel. They arrived in Wichita on Thanksgiving Day and ate a meal of turkey with Nancy's father, Curtis Dunn, who was living with Nancy's brother.

Nancy's father was dying of cancer. He had gone through life fearing that he might come down with cancer—he once took to his bed for eight months while claiming he had cancer when, in fact, he did not—and now he had come down with real cancer. He had lost a lot of weight that fall. He was like a human skeleton, down to less than a hundred pounds, but he was still a relatively young man, and his hair was black and curly and oiled with Vitalis. He looked so terrible that the children were afraid of him. He did his best to show sympathy for Jerry. "How awful it was, what happened to you Jaaxes,'' he said to Jerry. Jerry did not want to talk about it.

Nancy's father sat and slept in a reclining chair most of the day. At night, he couldn't sleep on account of the pain, and he would wake up at three o'clock in the morning, and get out of bed, and rummage around the house, looking for something. He smoked cigarettes continually, and complained that he couldn't taste his food, that he had lost his appetite. Nancy felt sorry for him, but she felt a distance from him that she could not overcome. He was a man of strong opinions, and lately, from the way he had been talking while he wandered about the house at night, it seemed that he was going to try to sell the family farm in Kansas and use the money to get himself to Mexico for a cure involving peach pits. Nancy was angry with him for having such ideas, and that anger was mixed with pity for him in his illness.

After they had finished their turkey with Nancy's father, they drove out to Andale, Kansas, a town northwest of Wichita, and ate another dinner, with Jerry's mother, Ada, and the rest of the Jaax family in Ada's house on the edge of town, near the grain elevator. Ada was a widow who lived alone in a ranch house that looked out across beautiful wheat fields. The fields were bare and planted with winter wheat, and Ada sat in her chair in the living room and stared outdoors. She could not watch television because she was afraid she would see a gun. They sat around the living room and talked, telling stories about the old days on Ada's farm, laughing and joking and trying to have a good time, and suddenly John's name would come up. The conversation would flag into silence, and everyone would look at the floor, not knowing what to say, and someone would start crying, and then they would see tears running down Ada's face. She had always been a strong woman, and none of her children had ever seen her cry. When she felt she could not stop it, she would get up and leave the room, and go into her bedroom and close the door.

They set up tables in the kitchen and served roast beef—the Jaaxes did not like turkey. After a while, people drifted into the living room with plates in their hands and watched a football game. The women, including Nancy, cleaned up around the kitchen and helped with the children. Afterward, Nancy and Jerry stayed on in Wichita for a few days to help Nancy's father get to the hospital for his cancer treatments. Then they drove back to Maryland in the van with their children.

Dan Dalgard spent an uneasy Thanksgiving week. On Monday, he called Peter Jahrling at the Institute to find out if Jahrling had any further news about what had been killing the monkeys at Reston. Jahrling now had a tentative diagnosis. It looked like the animals really did have SHF. Bad for monkeys, no problem for humans. He said to Dalgard that he felt strongly that it was simian fever, but he was reluctant to say so categorically. He wanted to play it carefully until the final tests were finished.

Dalgard hung up the phone believing that his decision to sacrifice the monkeys in Room F had been correct. Those monkeys had been infected with simian fever and would have died anyway. What now worried Dalgard was the possibility that the virus had somehow escaped from Room F. It might be quietly working its way through the building, in which case monkeys might start dying in other rooms. And then the virus would be very hard to control.

On Thanksgiving morning, Dan and his wife drove to Pittsburgh, to be with Dan's wife's parents. They drove back to Virginia on Friday, and Dan headed over to the monkey house to see if there had been any changes. He was shocked by what he found. Over Thanksgiving, five monkeys had died in Room H, two doors down the hall from Room F. So the virus was moving, and what was worse, it was skipping rooms as it moved. How could it do that? Five dead monkeys in one room during the night. . . . He felt very uneasy.

Medusa

Early on Monday morning of the week following Thanksgiving, Tom Geisbert went to work at the Institute wearing blue jeans, a flannel shirt, and cowboy boots, as a kind of memento of his time in the woods. He was anxious to check up on the button of dead monkey cells that he had harvested from the little flask just before he had gone hunting. He wanted to look at the cells in his electron microscope to try to find some visual evidence that they were infected with simian fever.

The button was a dot the size of a toast crumb, embedded in a tiny plug of yellow plastic. He unlocked a filing cabinet and removed his diamond knife. A diamond knife is a metal object no larger than a small pocket-size pencil sharpener—about an inch long. It costs about four thousand dollars. It has a diamond edge—a large, flawless prism-shaped diamond, a gem-quality stone.

He carried the diamond knife and the plug of plastic containing the toast crumb of cells into the cutting room. He sat down at a table, facing the cutting machine, and fitted his diamond knife into it, taking extreme care not to let his fingers touch the edge of the knife. One touch of a fingertip would destroy it. The diamond would also cut your fingertip, perhaps badly. The knife is extraordinarily sharp. It has the sharpest cutting edge of any tool on earth. It is sharp enough to split a virus cleanly in half, like a razor blade going through a peanut. If you consider the idea that a hundred million viruses could cover the dot on this *i*, then you get an idea of the sharpness of a diamond knife. If you happened to cut yourself with it, it would go through your skin without resistance, as if your skin were air—and it would split individual blood cells as it went through your finger. And then the knife edge would be covered with skin oil and blood cells, and would be ruined.

Tom looked into the eyepieces of a microscope that was attached to the cutting machine. Now he could see the toast crumb clearly. He threw a switch, and the machine hummed, and the sample began to move back and forth, the toast crumb sliding across the edge of the diamond knife. The cutting machine worked like a deli slicer, peeling off slices about this size:

The slices fell onto a droplet of water and rested on the droplet's surface. Each contained as many as ten thousand cells, and the cells themselves were split by the knife. The blade peeled off slice after slice. They spread out like lily pads.

He took his eyes away from the microscope and looked around the table until he found a wooden stick that had a human eyelash glued to it with a dab of nail polish. It was a device for handling the slices. The eyelash had come from one of the women in the lab—it was generally believed that she had superior eyelashes for this kind of work, not too thick and not too thin, tapered, ending in fine points. He poked the eyelash into the water droplet and stirred it, separating the slices from one another. With the tip of the eyelash, he then lifted a few damaged slices out of the water and wiped them on a piece of tissue paper to get rid of them.

Next, using a pair of tweezers, he picked up a small metal grid. The grid was this size—•—and it was made of copper. Holding the grid with his tweezers, he dipped it into the water and brought it up slowly underneath a floating slice, like a fisherman lifting up a dip net. The slice was now stuck to the grid. Still holding the grid with his tweezers, he put it into a tiny box. He carried the box down the hall to a darkened room. In the middle of the room stood a metal tower taller than a person. This was his electron microscope. My scope, he thought; he was very fond of it. He opened the tiny box, lifted out the grid with tweezers, and fitted it into a steel rod the size of a tire iron—the sample holder, it was called. He slid the rod into the microscope until it clanked, locked in place. Now the slice, sitting on the grid, which was held in place by the tire iron, was positioned in the microscope, centered in the beam of electrons.

He switched off the lights in the room and sat down at a console that was covered with dials and digital readouts. In the middle of the console there was a viewing screen. The room had become the command deck of a starship, and the viewing screen was a window that looked into the infinity within.

He hit a switch, hunched down in his chair, and put his head close to the viewing screen. His face glowed greenish in the light of the screen, and was reflected in the glass: long hair, serious expression, deep-set eyes that scanned the terrain. He was looking into a corner of one cell. It was like looking at a landscape from high altitude. It was a cellscape. What loomed before his eyes was a huge, complicated vista, crowded with more detail than the mind could absorb. You could spend days scanning cells, looking for a virus. In one slice, there might be thousands of cells that needed to be searched—and you still might not find what you were looking for. The incredible thing about living systems is that no matter how small the view, it is just as complicated as ever. He could see forms and shapes that resembled rivers and streams and oxbow lakes, and he could see specks that might be towns, and he could see belts of forest. It was an aerial view of rain forest. The cell was a world down there, and somewhere in that jungle hid a virus.

He turned a knob, and the cellscape drifted across his field of view, and he wandered through it. He zoomed in. The scene rushed up toward him.

His breath stopped. Wait a minute—there was something wrong with this cell. This cell was a mess. It wasn't just dead—it had been destroyed. It was blown apart. And it was crawling with worms. The cell was wall-to-wall with worms. Some parts of the cell were so thick with virus they looked like buckets of rope. There was only one kind of virus that looked like rope. A filovirus.

He thought, *Marburg.* Oh, no. This stuff looks like Marburg. He hunched over the screen. Then his stomach screwed up into a knot and turned over, and he felt an unpleasant sensation. The puke factor. He almost panicked, almost ran out of the room shouting, ''Marburg! We've got Marburg!'' He thought, Is this really happening? He sucked in his breath. He didn't know if this thing was Marburg, but it sure as hell looked like a filovirus, a thread virus. Then an image came into his mind—an image of Peter Cardinal's liver cells exploded and flooded with snakes. He brought the image into mental focus and compared it with what he saw on the screen. He knew exactly what the Cardinal strain looked like because he had memorized its curlicues and Cheerio shapes. What the virus did to that boy . . . the devastating effect on that boy's tissues . . . oh, man!—oh, man!—Pete and I smelled this stuff. Pete and I have been handling this stuff, and this is a Biosafety Level 4 agent. Marburg . . . oh, man . . . A foul feeling washed over him, a sudden awareness of male reproductive glands hanging on the exterior of the body between the legs . . . testicles the size of pears, black and putrid, the skin peeling off them.

He began snapping photographs with his microscope.

Several negatives came out of the machine. He carried them into a darkroom and switched out the lights and began developing them. In pitch-darkness, he had time to think. He counted the days back to the date of his exposure. Let's see, he had sniffed that flask on the Friday before he went hunting. That would have been . . . ten days ago. What's the incubation period for Marburg? He didn't know offhand. Let's see—monkeys that inhaled Marburg virus took a long time to develop the disease, from six to eighteen days. He was on day ten.

I am in the window to be sick. I am in prime time to be dropping over! Did I have a headache yesterday? Do I have a headache now? Do I have a fever? He placed his hand on his forehead. Feels okay. Just because I don't get a headache on day ten doesn't mean I won't get a headache on day twelve. How deep did I breathe when I sniffed that flask? Did I snap the cap? That would spray stuff around. I can't remember. Did I rub my eye with my finger afterward? I can't remember. Did I touch my mouth with my finger? I might have, I don't know.

He wondered if he had made a mistake. Maybe this virus wasn't Marburg. He was only an intern; he was just learning this stuff. Finding major Biosafety Level 4 agents on the outskirts of Washington, D.C., is not the kind of thing interns do every day. Maybe this isn't a filovirus. How sure am I? If you go and tell your boss that you've found Marburg virus and you are wrong, your career goes down the tubes. If you make a bad call, then first of all you start a panic. Second, you become a laughingstock.

He switched on the darkroom light and pulled the negatives out of the bath and held them up to the light.

He saw virus particles shaped like snakes, in negative images. They were white cobras tangled among themselves, like the hair of Medusa. They were the face of Nature herself, the obscene goddess revealed naked. This life form thing was breathtakingly beautiful. As he stared at it, he found himself being pulled out of the human world into a world where moral boundaries blur and finally dissolve completely. He was lost in wonder and admiration, even though he knew that he was the prey. Too bad he couldn't bring it down with a clean shot from a rifle.

He saw something else in the pictures that left him frightened and filled with awe. The virus had altered the structure of the cell almost beyond recognition. It had transformed the cell into something that resembled a chocolate-chip cookie that was mostly chocolate chips. The ''chips'' were crystal-like blocks of pure virus. He knew them as ''inclusion bodies.''

They were broods of virus getting ready to hatch. As the virus grows inside a cell, crystalloids, or bricks, appear at the center. Then they move outward, toward the surface of the cell. When a brick touches the inner surface of the cell wall, it breaks apart into hundreds of individual viruses. The viruses are shaped like threads. The threads push through the cell wall and grow out of the cell, like grass rising from seeded loam. As the bricks appear and move outward, they distort the cell, causing it to bulge and change shape, and finally the cell pops—it bursts and dies. The threads break away from the cell and drift into the bloodstream of the host, multiplying and taking over more cells and forming bricks and bursting the cells.

As he looked at the bricks, he realized that what he had thought was ''pepper'' when he had looked at the cells in the flask ten days ago—those specks in the cells—were really inclusion bodies. That was also why the cells had looked swollen and fat. Because they were pregnant and jammed with bricks of virus. Because they were getting ready to burst.

The First Angel

Tom Geisbert printed the negatives on eight-by-ten glossy paper and headed for the office of his boss, Peter Jahrling. He carried his photographs down a long hallway, went downstairs and through a security door, swiping his ID card across a sensor, and entered a warren of rooms. He nodded to a soldier—there were soldiers everywhere, going about their business at USAMRIID—and went up another flight of stairs and past a conference room that displayed a map of the world on the wall. In this room, Army doctors and officials discussed outbreaks of virus. A meeting was in progress in the room. Beyond it, he came to a cluster of offices. One of them was an awe-inspiring mess, papers everywhere. It belonged to Gene Johnson, the biohazard expert who had led the expedition to Kitum Cave. Across the way was Peter Jahrling's office. It was neatly kept and small, but it had a window. Jahrling had placed his desk under the window to get some extra light. On the walls he had hung drawings done by his children. There was a drawing by his daughter that showed a rabbit under a shining yellow sun. A shelf held an African sculpture of a human hand holding an egg on the tips of its fingers, as if the egg contained something interesting about to hatch.

"What's up, Tom?" Jahrling asked.

"We have a big problem here," Geisbert said, and he placed the photographs in a row on Jahrling's desk. It was a gray November day, and the light from the window fell gently on the images of Medusa. "This came from the Reston monkeys," Geisbert said. "I think it's a filovirus, and there may be a good chance it's Marburg."

Jahrling remembered sniffing the flask and said, "You're playing a joke on me. This isn't funny."

"This is no joke, Pete."

"Are you sure?" Jahrling asked.

Geisbert said he felt very sure.

Jahrling looked carefully at the photographs. Yes, he could see worms. Yes, he and Geisbert might have breathed it into their lungs. Well, they didn't have headaches yet. He remembered remarking to the pathologist, as he cut up the little pink chunk of mystery meat in the tin foil, *"Good thing this ain't Marburg."* Yeah, right.

"Is this stuff the right size?" Jahrling asked. He got a ruler and measured the particles.

"It looks a little long to be Marburg," Geisbert said. Marburg particles form loops like Cheerios. This stuff was more like spaghetti. They opened a textbook and compared Geisbert's pictures with the textbook pictures.

"It looks good to me," Jahrling said. "I'm going to show it to C. J. Peters."

Jahrling, a civilian, had decided to notify the military chain of command. It started with Colonel Clarence James Peters, MD. He was the chief of the disease-assessment division at the Institute, the doctor who dealt with the dangerous unknowns. ("The interesting stuff," as he called it.) C. J. Peters had built up this division almost singlehandedly, and he ran it singlehandedly. He was a strange sort of military man, easygoing and casually brilliant. He had wire-rimmed glasses, a round, ruddy, pleasant face with a mustache, a light Texas drawl. He was not a large man, but he liked to eat, and he believed himself to be overweight. He spoke fluent Spanish, which he had learned during his years in the jungles of Central and South America, hunting for hot agents. He was required by Army regulations to show up for work at eight o'clock in the morning, but he usually drifted in around ten o'clock, and then worked until all hours of the night. He disliked wearing a uniform. Usually he wore faded blue jeans with a flaming Hawaiian shirt, along with sandals and dweebish white socks, looking like he had just spent the night in a Mexican hotel. His excuse for his lack of uniform was that he suffered from athlete's foot, an incurable tropical strain that he'd picked up in Central America and could never quite get rid of, and so he had to wear socks with sandals in order to keep air circulating around his toes—and the jeans and the flaming shirt were part of the package.

C. J. Peters could swim through a bureaucracy like a shark. He inspired great loyalty in his staff, and he made enemies easily and deliberately, when it suited him or the needs of his staff. He drove a red Toyota that had seen better days. On his travels in rain forests and tropical savannas, he ate with pleasure whatever the locals were eating. He had consumed frogs, snakes, zebra meat, jellyfish, lizards, and toads cooked whole in their skin, but he thought he had never eaten salamanders, at least none that he had been able to identify in a soup. He had eaten boiled monkey thigh, and he had drunk banana beer fermented with human saliva. In central Africa, while leading an expedition in search of Ebola virus, he had found himself in termite country during swarming season, and he had waited by termite nests and collected the termites as they swarmed out and had eaten them raw. He thought they had a nice sort of nutty taste. He liked termites so much that he refrigerated them with his blood samples, to keep the termites fresh all day so that he could snack on them like peanuts with his evening gin as the sun went down over the African plains. He was fond of suffocated guinea pig baked in its own blood and viscera. The guinea pig is split open like a book, offering treasures, and he enjoyed picking out and eating the guinea pig's lungs, adrenal glands, and brain. And then, inevitably, he would pay a price. "I always get sick, but it's worth it," he once said to me. He was a great believer in maps, and his offices always contained many maps hung on the walls, showing locations of outbreaks of virus.

Jahrling put Geisbert's photographs in a folder. He didn't want anyone to see them. He found Peters at a meeting in the conference room that held the map of the world. Jahrling tapped him on the shoulder. "I don't know what you are doing right now, C. J., but I've got something more important."

"What is it?"

Jahrling held the folder closed. "It's a little sensitive. I really don't want to flash it here."

"What's so sensitive?"

Jahrling opened the folder slightly, just enough to give C. J. a glimpse of spaghetti, and snapped it shut.

The colonel's face took on a look of surprise. He stood up, and without a word to the others, without even excusing himself, he walked out of the room with Jahrling. They went back to Jahrling's office and closed the door behind them. Geisbert was there, waiting for them.

Jahrling spread the photographs on his desk. "Take a look at these, C. J."

The colonel flipped through the photographs. "What's this from, anyway?" he asked.

"It's from those monkeys in Reston. It doesn't look good to me. Tom thinks it's Marburg."

"We've been fooled before," C. J. said. "A lot of things look like worms." He stared at the photographs. The worms were unmistakable—and there were the crystalloids—the bricks. It looked real. It felt real. He experienced what he would later describe as "a major pucker factor" setting in.

(This is a military slang term that refers to a certain tightening sensation in the nether regions of the body, in response to fear.) He thought, This is going to be an awful problem for that town in Virginia and those people there. "The first question," he went on, "is what are the chances of laboratory contamination?" The stuff could be the Army's own Cardinal strain—it might have somehow leaked out of a freezer and gotten into those flasks. But that seemed impossible. And the more they pondered it, the more impossible it seemed. The Cardinal strain was kept in a different area of the building, behind several walls of biocontainment, a long distance from the monkey flasks. There were multiple safeguards to prevent the accidental release of a virus like Marburg Cardinal. That just wasn't possible. It could not be a contamination. But it might be something other than a virus. It might be a false alarm.

"People around here see something long and stringy, and they think they've got a filovirus," C. J. Peters said. "I'm skeptical. A lot of things look like Marburg."

"I agree," Jahrling replied. "It could be nothing. It could be just another Loch Ness monster."

"What are you doing to confirm it?" the colonel asked him.

Jahrling explained that he was planning to test the cells with human blood samples that would make them glow if they were infected with Marburg.

"Okay, you're testing for Marburg," C. J. said. "Are you going to include a test for Ebola?"

"Sure. I already thought of that."

"When will your tests be done? Because if those monkeys have Marburg, we have to figure out what to do."

Dan Dalgard, for example, was a prime candidate for coming down with Marburg, because he had dissected that monkey.

"I'll have a definite yes or no on Marburg by tomorrow," Jahrling said.

C. J. Peters turned to Tom Geisbert and said that he wanted more proof—he wanted pictures of the agent actually growing in monkey liver from a monkey that had died in the monkey house. That would prove that it lived in the monkeys.

C. J. could see that a military and political crisis was brewing. If the public found out what Marburg does, there could be panic. He stood up with a photograph of snakes in his hand and said, "If we are going to announce that Marburg has broken out near Washington, we had better be damned sure we are right." Then he dropped the photograph on Jahrling's desk and returned to his meeting under the map of the world.

After C. J. Peters left Jahrling's office, a delicate conversation occurred between Peter Jahrling and Tom Geisbert. They shut the door and talked quietly about the whiffing incident. It was something they had better get straight between them. Neither of them had mentioned to Colonel C. J. Peters that they had whiffed that flask.

They counted the days back to their exposure. Ten days had passed since they had uncapped the flask and breathed what could be eau de Marburg. Tomorrow would be day eleven. The clock was ticking. They were in the incubation period. What were they going to do? What about their families?

They wondered what Colonel Peters would do if he found out what they had done. He might order them into the Slammer—the Level 4 biocontainment hospital. They could end up in the Slammer behind air locks and double steel doors, tended by nurses and doctors wearing space suits. A month in the Slammer while the doctors hovered over you in space suits drawing samples of your blood, just waiting for you to crash.

The doors of the Slammer are kept locked, the air is kept under negative pressure, and your telephone calls are monitored—because people have emotional breakdowns in the Slammer and try to escape. They start flaking out by the second week. They become clinically depressed. Noncommunicative. They stare at the walls, speechless, passive, won't even watch television. Some of them become agitated and fearful. Some of them need to have a continual drip of Valium in the arm to keep them from pounding on the walls, smashing the viewing windows, tearing up the medical equipment. They sit on death row in solitary confinement, waiting for the spiking fevers, horrible pain in the internal organs, brain strokes, and finally the endgame, with its sudden, surprising, uncontrollable gushes of blood. Most of them claim loudly that they have not been exposed to anything. They deny that anything could go wrong with them, and ordinarily nothing does go wrong with them, physically, in the Slammer, and they come out healthy. Their minds are another story. In the Slammer, they become paranoid, convinced that the Army bureaucracy has forgotten about them, has left them to rot. When they come out, they are disoriented. They emerge through the air-lock door, pale, shaken, tentative, trembling, angry with the Army, angry with themselves. The nurses, trying to cheer them up, give them a cake studded with the number of candles equal to the number of days they've been living in the Slammer. They blink in confusion and terror at a mass of flaming candles on their Slammer cake, perhaps more candles than they've ever seen on one of their own birthday cakes. One guy was locked in the Slammer for forty-two days. Forty-two candles on his Slammer cake.

Many people who have been isolated in the Slammer choose to cut down on their work in Level 4, begin to find all kinds of excuses for why they really can't put on a space suit today or tomorrow or the day after that. Many of the people who have been in the Slammer end up quitting their jobs and leaving the Institute altogether.

Peter Jahrling felt that, on the whole, he was not at much

risk of contracting the virus, nor was Tom. If he did contract it, he would know soon enough. His blood would test positive, or he would get a headache that wouldn't go away. In any case, he believed very strongly that Marburg wasn't easy to catch, and he didn't think there was any danger to his family or to anyone else around town.

But think about Dan Dalgard cutting into monkeys. Bending over and *breathing* monkey when he opened their abdomens. He was bending over their intestines, inhaling fumes from a pool of Marburg blood. So then, why isn't Dalgard dead? Well, he reasoned, nothing's happened to Dalgard, so maybe nothing will happen to us.

Where had it come from? Was it a new strain? What was it capable of doing to humans? The discoverer of a new strain of virus gets to name it. Jahrling thought about that, too. If he and Tom were locked up in the Slammer, they would not be able to carry out any research on this virus. They were on the verge of a major discovery, and the glory of it perhaps tantalized them. To find a filovirus near Washington was the discovery of a lifetime.

For all these reasons, they decided to keep their mouths shut.

They decided to test their blood for the virus. Jahrling said to Geisbert, "We are going to get blood samples drawn from ourselves *like right now.*" If their blood went positive, they could immediately report to the Slammer. If their blood remained negative and they didn't develop other symptoms, then there was little chance they could infect anyone else.

Obviously they did not want to go to the regular clinic to have an Army nurse take their blood: that would be a tip-off to the military authorities that they thought they'd been exposed. So they found a friendly civilian technician and he twisted a rubber band around their arms, and they watched while he filled some tubes with their blood. He understood what had happened, and he said he would keep his mouth shut. Jahrling then put on a space suit and carried his own blood into his Level 4 hot lab. He also took with him Geisbert's blood and the flasks of milky stuff. It was very strange, handling your own blood while wearing a space suit. It seemed, however, quite risky to let his blood lie around where someone might be accidentally exposed to it. His blood *had* to be biocontained in a hot zone. If it was infected with Marburg, he didn't want to be responsible for it killing anyone. He said to himself, Given that this was a piece of mystery meat sniffed out of a monkey carcass, I should have been a little more careful. . . .

Meanwhile, Tom Geisbert went off to collect some pickled monkey liver that he could photograph for viruses, hoping to prove that the Marburg-like agent lived in the monkeys. He found a plastic jug that contained sterilized pieces of liver from Monkey O53. He fished some liver out of the jug, clipped a few bits off it, and fixed the bits in plastic. This was a slow job and took many hours to finish. He left the plastic to cure overnight and went home for a couple of hours to try to get some sleep.

The Second Angel

NOVEMBER 28, TUESDAY

Tom Geisbert lived in a small town in West Virginia, across the Potomac River from Maryland. After his separation from his wife, his two children had stayed with her for a time, and now they were staying with him, or rather, they were staying with his parents in their house down the road. Both his children were toddlers.

He got up at four o'clock in the morning, drank a cup of coffee, and skipped breakfast. He drove his Bronco in pitch darkness across the Potomac and through Antietam National Battlefield, a broad ridge of cornfields and farmland scattered with stone monuments to the dead. He passed through the front gate of Fort Detrick, parked, and went past the security desk and into his microscope area.

The dawn came gray, gusty, and warm. As light glimmered around the Institute, Tom sliced pieces of monkey liver with his diamond knife and put them into the electron microscope. A few minutes later, he took a photograph of virus particles budding directly out of cells in the liver of Monkey O53. These photographs were definite proof that the virus was multiplying in the Reston monkeys—that it was not a laboratory contamination. He also found inclusion bodies inside the monkey's liver cells. The animal's liver was being transformed into crystal bricks.

He carried his new photographs to Peter Jahrling's office. Then they both went to see Colonel C. J. Peters. The colonel stared at the photographs. Okay—he was convinced, too. The agent was growing in those monkeys. Now they would have to wait for Jahrling's test results, because that would be the final confirmation that it was indeed Marburg.

Jahrling wanted to nail down this Marburg as fast as he could. He spent most of the day in a space suit, working in his hot lab, putting together his tests. In the middle of the day, he decided that he had to call Dan Dalgard. He couldn't wait any longer, even without test results. He wanted to warn Dalgard of the danger, yet he wanted to deliver the warning carefully, so as not to cause a panic. "You definitely have SHF in the monkey house," he said. "We have definitely confirmed that. However, there is also the possibility of a *second agent* in at least some of the animals."

"What agent? Can you tell me what agent?" Dalgard asked.

"I don't want to identify the agent right now," Jahrling said, "because I don't want to start a panic. But there are serious potential public health hazards associated with it, if, in fact, we are dealing with this particular agent."

Somehow, the way Jahrling used the words *panic* and *particular* made Dalgard think of Marburg virus. Everyone who handled monkeys knew about Marburg. It was a virus that could easily make people panic.

"Is it Marburg or some similar agent?" Dalgard asked.

"Yes, something like that," Jahrling said. "We'll have confirmation later in the day. I'm working on the tests now. I feel it's *unlikely* the results will be positive for this second agent. But you should take precautions not to do any necropsies on any animals until we've completed the tests. Look, I don't want to set off too many whistles and bells, but I don't want you and your employees walking into that room unnecessarily.''

''How soon can you get back to me with a definite yes or no about this second agent? We need to know as soon as possible.''

''I'll call you back today. I promise,'' Jahrling said.

Dalgard hung up the phone highly disturbed, but he maintained his usual calm manner. So there was a second agent in the building, and it sounded as if it was Marburg. The people who had died in Germany, he knew, had been handling raw, bloody monkey meat. The meat was full of virus, and they got it on their hands, or they rubbed it on their eyelids. He and other people at the company had been cutting into sick monkeys since October—and yet no one had become sick. Everyone had worn rubber gloves. He wasn't afraid for himself—he felt fine—but he began to worry about the others. He thought, Even if the virus is Marburg, the situation is still no different from before. We're still stuck in a pot. The question is how to get ourselves out of this pot. He called Bill Volt and ordered him not to cut into any more monkeys. Then he sat in his office, getting more and more annoyed as the day darkened and Peter Jahrling did not call him back. He wondered if any of the men had cut themselves with a scalpel while performing a dissection of a diseased monkey. Chances were they wouldn't file an accident report. He knew for sure that he had not cut himself. But he had performed a mass sacrifice of approximately fifty animals. He had been in contact with the blood and secretions of *fifty* animals. That had been on the sixteenth of November. Eleven days ago. He should be showing some symptoms by now. Bloody nose, fever, something like that. Or maybe he just hadn't broken with virus yet.

At five-thirty, he called Jahrling's office and got a soldier on the phone, who answered by saying, ''How can I help you, sir or ma'am? . . . I'm sorry, sir, Dr. Jahrling is not in his office. . . . No sir, I don't know where he is, sir. . . .

No, he has not left work. May I take a message, sir?'' Dalgard left a message for Jahrling to call him at home. He was feeling steadily more annoyed.

1500 HOURS

While Dalgard fretted, Jahrling was in his space suit. He worked steadily all afternoon in his own lab, hot zone AA-4, at the center of the building, where he fiddled with the flasks of virus culture from the monkey house. It was a slow, irritating job. His tests involved making the samples glow under ultraviolet light. If he could make the samples glow, then he knew he had the virus.

In order to do this, he needed to use blood serum from human victims. The blood serum would react to viruses. He went to the freezers, and got out vials of frozen blood serum from three people. Two of the people had died; one had survived. They were:

1. *Musoke.* A test for Marburg. Serum from the blood of Dr. Shem Musoke, a survivor. (Presumably reactive against the Kitum Cave strain, which had started with Charles Monet and jumped into Dr. Musoke's eyes in the black vomit.)
2. *Boniface.* A test for Ebola Sudan. From a man named Boniface who died in Sudan.
3. *Mayinga.* A test for Ebola Zaire. Nurse Mayinga's blood serum.

The test was delicate, and took hours to complete. It was not made easier by the fact that he was shuffling around in his space suit the whole time. First he put droplets of cells from the monkey culture onto glass slides, and let them dry, and treated them with chemicals. Then he put drops of the blood serum on the slides. The blood would glow in the presence of the target virus.

Now it was time to look. This had to be done in total darkness, because the glow would be faint. He shuffled over to a storage closet, and went inside it, and closed the door

behind him. A microscope sat on a table in the closet, and
there was a chair, and from the wall hung an air hose. He
plugged the hose into his space suit and put the slides into
the microscope. Then he turned out the lights. He felt around
in the darkness for the chair, and sat down. This was not a
fun place to be if you happened to have a touch of claustrophobia—sitting in a pitch-black Level 4 closet while wearing
a space suit. Peter Jahrling had made his peace with suffoca-
tion and darkness a long time ago. He waited for a minute to
give his eyes time to adapt to the dark, and the little sparkles
of light in his eyes as they adjusted to the darkness eventually faded away, while cool, dry air roared around his face
and whiffled the hair on his forehead. Then he looked
through the binocular eyepieces of the microscope. He wore
his eyeglasses inside his space suit, and that made it particularly difficult to see. He pressed the faceplate against his
nose and squinted. He moved his face from side to side. His
nose left a greasy streak inside his faceplate. He twisted his
helmet until it was turned nearly sideways. Finally he saw
through the eyepieces.

Two circles drifted into his sight, and he focused his eyes,
bringing the circles together. He was looking down into vast
terrain. He saw cells dimly outlined in a faint glow. It was
like flying over a country at night, over thinly populated
lands. It was normal to see a faint glow. He was looking for
a bright glow. He was looking for a city. He scanned the
slides with his eyes, back and forth, back and forth, moving
across the microscopic world, looking for a telltale greenish
glow.

The Musoke did not glow.

The Boniface glowed weakly.

To his horror, the Mayinga glowed brightly.

He jerked his head back. Aw, no! He adjusted his helmet
and looked again. The Mayinga blood serum was still glowing. The dead woman's blood was reacting to the virus in
the monkey house. He got an ugly feeling in the pit of his

stomach. Those monkeys didn't have Marburg. They had Ebola. *Those animals were dying of Ebola Zaire.* His stomach lurched and turned over, and he sat frozen in the dark closet, with only the sound of his air and the thud of his heart.

Chain of Command

This can't be Ebola Zaire, Peter Jahrling thought. Somebody must have switched the samples by accident. He looked again. Yeah, the Mayinga blood serum was definitely glowing. It meant he and Tom could be infected with Ebola Zaire, which kills nine out of ten victims. He decided that he had made a mistake in his experiment. He must have accidentally switched around his samples or gotten something mixed up.

He decided to do the test again. He turned on the lights in the closet and shuffled out into his lab, this time keeping careful track of his vials, bottles, and slides to make sure that nothing got mixed up. Then he carried the new samples back into the closet and turned out the lights and looked again into his microscope.

Once again, the Mayinga blood glowed.

So maybe it really was Ebola Zaire or something closely related to it—the dead woman's blood "knew" this virus, and reacted to it. Good thing this ain't Marburg—well, guess what, *it ain't Marburg*. This is the honker from Zaire, or maybe its twin sister. Ebola had never been seen outside Africa. What was it doing near Washington? How in the hell had it gotten here? What would it do? He thought, I'm onto something really hot.

He was wearing his space suit, but he didn't want to take the time to decon out through the air lock. There was an emergency telephone on the wall in his lab. He disconnected his air hose to extinguish the roar of air so that he could hear through the receiver, and he punched Colonel C. J. Peter's phone number.

"C. J.!" he shouted through his helmet. "IT'S PETE JAHRLING. IT'S REAL, AND IT'S EBOLA."

"Naw!" C. J. replied.

"YEAH."

"Ebola? It's gotta be a contamination," C. J. said.

"NO, IT ISN'T A CONTAMINATION."

"Could you have gotten your samples mixed up?"

"YEAH, I KNOW—MY FIRST THOUGHT WAS THAT SOMEBODY HAD SWITCHED THE SAMPLES. BUT THEY WEREN'T SWITCHED, C. J.—BECAUSE I DID THE TEST TWICE."

"Twice?"

"EBOLA ZAIRE BOTH TIMES. I'VE GOT THE RE-SULTS RIGHT HERE. I CAN PASS THEM TO YOU. TAKE A LOOK FOR YOURSELF."

"I'm coming down there," C. J. said. He hung up the phone and hurried downstairs to Jahrling's hot lab.

Jahrling, meanwhile, picked up a sheet of waterproof paper on which he had written down the results of his tests. He slid the paper into a tank full of EnviroChem. The tank went through the wall to a Level 0 corridor outside the hot zone. The tank worked on the same principle as a sliding cash drawer in a teller's window. You could pass an object from the hot zone through the tank into the normal world. The object would be disinfected on its way through the tank.

C. J. stood at a thick glass window on the other side, looking in at Jahrling. They waited for several minutes while the chemicals penetrated the paper and sterilized it. Then C. J. opened the tank from his side and removed the paper, dripping with chemicals, and held it in his hands. He motioned to Jahrling through the window: *Go back to the phone.*

Jahrling shuffled back to the emergency telephone and waited for it to ring. It rang, and there was C. J.'s voice on the line: "Get out of there, and let's go see the commander!"

It was time to move this thing up the chain of command.

Jahrling deconned out through the air lock, got dressed in his street clothes, and hurried to C. J. Peters's office, and they both went to the office of the commander of USAMRIID, a colonel named David Huxsoll. They brushed past his secretary—told her it was an emergency—and sat down at a conference table in his office.

"Guess what?" C. J. said. "It looks like we've found a filovirus in a bunch of monkeys outside Washington. We've recovered what we think is Ebola."

Colonel David Huxsoll was an expert in biohazards, and this was the sort of situation he thought the Institute was prepared to handle. Within minutes, he had telephoned Major General Philip K. Russell, MD, who was the commander of the United States Army Medical Research and Development Command, which has authority over USAMRIID, and had set up a meeting in Russell's office in another building at Fort Detrick.

Colonels Huxsoll and Peters spent a few moments talking about who else should be brought in. They hit upon Lieutenant Colonel Nancy Jaax, the Institute's chief of pathology. She could identify the signs of Ebola in a monkey. Huxsoll picked up his phone. "Nancy, it's Dave Huxsoll. Can you get over to Phil Russell's office right now? It's damned important."

It was a dark November evening, and the base was beginning to quiet down for the night. At the moment of sundown that day, there was no sun visible, only a dying of the light behind clouds that flowed off Catoctin Mountain. Jaax met Jahrling and the two colonels on their way across the parade ground beside the Institute. A detail of marching soldiers stopped before the flagpole. The group of people from the Institute also stopped. From a loudspeaker came a roar of a cannon and then the bugle music of "Retreat," cracklish and cheap-sounding in the air, and the soldiers lowered the flag while the officers came to attention and saluted.

C. J. Peters felt both embarrassed and oddly moved by the ceremony. "Retreat" ended, and the soldiers folded up the flag, and the Institute people continued on their way.

General Russell's office occupied a corner of a low-slung Second World War barracks that had been recently plastered with stucco in a hopeless effort to make it look new. It had a view of the legs of Fort Detrick's water tower. Consequently, the general never opened his curtains. The visitors sat on a couch and chairs, and the general settled behind his desk. He was a medical doctor who had hunted viruses in Southeast Asia. He was in his late fifties, a tall man with hair thinning on top and gray at the temples, lined cheeks, a long jaw, pale blue eyes that gave him a look of intensity, and a booming, deep voice.

C. J. Peters handed the general a folder containing the photographs of the life form that inhabited the monkey house.

General Russell stared. "Holy shit," he said. He drew a breath. "Man. That's a filovirus. Who the hell took this picture?" He flipped to the next one.

"These were done by my microscopist, Tom Geisbert," Jahrling said. "It could be Ebola. The tests are showing positive for Ebola Zaire."

C. J. then gave an overview of the situation, telling the general about the monkeys in Reston, and finishing with these words: "I'd say we have a major pucker factor about the virus in those monkeys."

"Well, how certain are you that it's Ebola?" General
Russell asked. "I'm wondering if this could be Marburg."

Jahrling explained why he didn't think it was Marburg.
He had done his test twice, he said, and both times the
samples were positive to the Mayinga strain of Ebola Zaire.
As he spoke to the general, he was very careful to say that
that test did not in itself prove that the virus was Ebola Zaire.
It showed only that is was closely related to Ebola Zaire. It
might be Ebola, or it might be something else—something
new and different.

C. J. said, "We have to be very concerned and very puckered if it is of the same ilk as Ebola."

They had to be very puckered, Russell agreed. "We have
a national emergency on our hands," he said. "This is an
infectious threat of major consequences." He remarked that
this type of virus had never been seen before in the United
States, and it was right outside Washington. "What the hell
are we going to do about it?" he said. Then he asked them if
there was any evidence that the virus could travel through
the air. That was a crucial question.

There *was* evidence, horrifying but incomplete, that Ebola
could travel through the air. Nancy Jaax described the incident in which her two healthy monkeys had died of presumably airborne Ebola in the weeks after the bloody-glove
incident, in 1983. There was more evidence, and she described that, too. In 1986, Gene Johnson had infected monkeys with Ebola and Marburg by letting them breathe it into
their lungs, and she had been the pathologist for that experiment. All of the monkeys exposed to airborne virus had died
except for one monkey, which managed to survive Marburg.
The virus, therefore, could infect the lungs on contact. Furthermore, the lethal airborne dose was fairly small: as small
as five hundred infectious virus particles. That many particles of airborne Ebola could easily hatch out of a single cell.
A tiny amount of airborne Ebola could nuke a building full
of people if it got into the air-conditioning system. The stuff

could be like plutonium. The stuff could be worse than plutonium because it could replicate.

C. J. said, "We *know* it's infectious by air, but we don't know *how* infectious."

Russell turned to Jaax and asked, "Has this been published? Did you publish it?"

"No, sir," she said.

He glared at her. She could see him thinking, Well, Jaax, why the hell *hasn't* it been published?

There were plenty of reasons, but she didn't feel like giving them just now. She believed that Gene Johnson, her collaborator, had difficulty writing papers. And, well, they just had not gotten around to publishing it, that was all. It happens. People sometimes just don't get around to publishing papers.

Hearing the discussion, Peter Jahrling chose not to mention to the general that he might have sniffed just a little bit of it. Anyway, he hadn't *sniffed* it, he had only *whiffed* it. He had kind of like waved his hand over it, just to bring the scent to his nose. He hadn't *inhaled* it. He hadn't like jammed the flask up into his nostril and snorted it or anything like that. Yet he had a feeling he knew what the general might do if he found out about it—the general would erupt in enough profanity to lift Jahrling off his feet and drop him into the Slammer.

Then there was the additional frightening possibility that this virus near Washington was not Ebola Zaire. That it was something else. Another hot strain from the rain forest. An unknown emerger. A new filovirus. And who could say how it moved or what it could do to humans? General Russell began to think out loud. "We could be in for a hellacious event," he said. "Given that we have an agent with a potential to cause severe human disease, and given that it appears to be uncontrolled in the monkey house, what do we do? We need to do the right thing, and we need to do it fast. How big is this sucker? And are people going to die?" He turned to

Colonel C. J. Peters and asked, "So what are our options here?"

C. J. had been thinking about this already. According to standard doctrine, there are basically three ways to stop a virus—vaccines, drugs, and biocontainment. There was no vaccine for Ebola. There was no drug treatment for Ebola. That left only biocontainment.

But how to achieve biocontainment? That was tricky. As far as C. J. could see, there were only two options here. The first option was to seal off the monkey colony and watch the monkeys die—and also keep a close watch on the people who had handled the monkeys and possibly put them into quarantine as well. The second option was to go into the building and sterilize the whole place. Kill the monkeys—give them lethal injections—burn their carcasses, and drench the entire building with chemicals and fumes—a major bio-hazard operation.

General Russell listened and said, "So option one is to cut the monkeys off from the rest of the world and let the virus run its course in them. And option two is to wipe them out. There aren't any more options."

Everyone agreed that there were no other options.

Nancy Jaax was thinking, It may be in the monkey house now, but it ain't going to stay there very damn long. She had never seen a monkey survive Ebola. And Ebola is a species jumper. All of those monkeys were going to die, and they were going to die in a way that was almost unimaginable. Very few people on earth had seen Ebola do its work on a primate, but she knew exactly what it could do. She did not see how the virus could be contained unless the monkey house was set up for quarantine with an independently filtered air supply. She said, "How ethical is it to let these animals go a long time before they die? And how do we assure the safety of people in the meantime? I've watched these animals die of Ebola, and it's not a fun way to go—

they're sick, sick, sick animals.'' She said that she wanted to
go into the monkey house to look at the monkeys. ''The
lesions are easy to miss unless you know what you are looking for,'' she said, ''and then it's as plain as the nose on your
face.''

She also wanted to go there to look at pieces of tissue
under a microscope. She wanted to look for crystalloids, or
''inclusion bodies.'' Bricks. If she could find them in the
monkey meat, that would be another confirmation that the
monkeys were hot.

Meanwhile, there was the larger question of politics.
Should the Army become involved? The Army has a mission, which is to defend the country against military threats.
Was this virus a military threat? The sense of the meeting
ran like this: military threat or no, if we are going to stop this
agent, we've got to throw everything at it that we've got.

That would create a small political problem. Actually it
would create a large political problem. The problem had to
do with the Centers for Disease Control in Atlanta, Georgia.
The C.D.C. is the federal agency that deals with emerging
diseases. It has a mandate from Congress to control human
disease. This is the C.D.C.'s lawful job. The Army does not
exactly have a mandate to fight viruses on American soil. Yet
the Army has the capability and the expertise to do it. Everyone in the room could see that a confrontation might boil up
with the C.D.C. if the Army decided to move in on the
monkey house. There were people at the C.D.C. who could
be jealous of their turf. ''The Army doesn't have the statu-
tory responsibility to take care of this situation,'' General
Russell pointed out, ''but the Army has the capability. The
C.D.C. doesn't have the capability. We have the muscle but
not the authority. The C.D.C. has the authority but not the
muscle. And there's going to be a pissing contest.''

In the opinion of General Russell, this was a job for
soldiers operating under a chain of command. There would

be a need for people trained in biohazard work. They would have to be young, without families, willing to risk their lives. They would have to know each other and be able to work in teams. They had to be ready to die.

In fact, the Army had never before organized a major field operation against a hot virus. The whole thing would have to be put together from scratch.

Obviously there were legal questions here. Lawyers were going to have to be consulted. Was this legal? Could the Army simply put together a biohazard SWAT team and move in on the monkey house? General Russell was afraid the Army's lawyers would tell him that it could not, and should not, be done, so he answered the legal doubts with these words: "A policy of moving out and doing it, and asking forgiveness afterward, is much better than a policy of asking permission and having it denied. You never ask a lawyer for permission to do something. We are going to do the needful, and the lawyers are going to tell us why it's legal."

By this time, the people in the room were shouting and interrupting one another. General Russell, still thinking out loud, boomed, "So the next question is, Who the fuck is going to pay for it?" Before anyone had a chance to speak, he answered the question himself. "I'll get the money. I'll beat it out of somebody."

More shouting.

The general's voice rose above the noise. "This is a big one coming, so let's not screw it up, fellas," he said. "Let's write the right game plan and then execute it." In the Army, an important job is called a mission, and a mission is always carried out by a team, and every team has a leader. "We have to agree on who is going to be in charge of this operation," the general continued. "C. J. Peters has got this action here. He's in charge of the operation. He's the designated team leader. Okay? Everybody agreed on that?"

Everybody agreed.

"C. J., what we need is a meeting," the general said.

"Tomorrow we're going to have a meeting. We have to call everybody."

He looked at the clock on the wall. It was five-thirty, rush hour. People were leaving work, monkeys were dying in Reston, and the virus was on the move. "We've got to pull the chain on this whole thing," the general said. "We'll have to inform everybody simultaneously, as soon as possible. I want to start with Fred Murphy at the C.D.C. I don't want him to be sandbagged by this."

Frederick A. Murphy was one of the original discoverers of Ebola virus, the wizard with an electron microscope who had first photographed the virus and whose work had hung in art museums. He was an old friend of General Russell's. He was also an important official at the C.D.C., the director of the National Center for Infectious Diseases.

Russell put his hand on the telephone on his desk. He stared around the room. "One last time: are you sure you've got what you think you've got? Because I'm gonna make this phone call. If you don't have a filovirus, we will look like real assholes."

Around the room, one by one, they told him they were convinced it was a thread virus.

"All right. Then I'm satisfied we've got it."

He dialed Murphy's number in Atlanta.

"Sorry—Dr. Murphy has gone home for the day."

He pulled out his black book and found Murphy's home phone number. He reached Murphy in his kitchen, where he was chatting with his wife. "Fred. It's Phil Russell. . . . Great, how about yourself? . . . Fred, we've isolated an Ebola-like agent outside Washington. . . . Yeah. Outside Washington."

A grin spread over Russell's face, and he held the phone away from his ear and looked around the room. Evidently his friend Murphy was having some kind of a noisy reaction. Then General Russell said into the receiver, "No, Fred, we're not smoking dope. We've got an Ebola-like virus.

We've seen it. . . . Yeah, we have pictures.'' There was a pause, and he put his hand over the mouthpiece and said to the room, "He thinks we've got crud in our scope."

Murphy wanted to know who took the pictures and who analyzed them.

"It was a kid who took the pictures. Young guy named— what's his name?—Geisbert. And we're looking at them right here."

Murphy said he would fly up to Fort Detrick tomorrow morning to look at the pictures and review the evidence. He took it extremely seriously.

1830 HOURS, TUESDAY

Dan Dalgard had to be called, and they had to notify Virginia state health authorities. "I'm not even sure who the state authorities are," Russell said. "And we've got to get them on the phone right now." People were leaving work. "We'll have to call people at home. It's going to be a bunch of phone calls." What county was that monkey house located in? Fairfax County, Virginia. My, oh my, a *nice* place to live. Fairfax County—beautiful neighborhoods, lakes, golf courses, expensive homes, good schools, and Ebola. "We'll have to call the county health department," the general said. They would also have to call the U.S. Department of Agriculture, which has control over imported monkeys. They would have to call the Environmental Protection Agency, which has jurisdiction in cases of environmental contamination by an extreme biohazard. General Russell also decided to call an assistant secretary of defense, just to get the Pentagon notified.

People left the room and fanned out along the hallways, going into empty offices and making the calls. C. J. Peters, now the team leader, went into another office down the hall and called Dan Dalgard's office, with Peter Jahrling on an extension line. Dalgard had gone home. They called Dalgard's home, and Dalgard's wife told him that Dan hadn't arrived yet. At about half past six, they called Dalgard's house again, and this time they got him. "This is Colonel C. J. Peters, up at USAMRIID. I'm the chief of the disease-assessment division. . . . How do you do? . . . Anyway, I'm calling to report that the second agent is apparently not Marburg. The second agent is Ebola virus."

"What is Ebola?" Dalgard asked. He had never heard of Ebola. The word had no meaning for him.

In his smoothest Texas voice, C. J. Peters said, "It's a rather rare viral disease that has been responsible for human fatalities in outbreaks in Zaire and Sudan within the past ten or twelve years."

Dalgard was starting to feel relieved—good thing it isn't Marburg. "What is the nature of Ebola virus?" he asked.

C. J. described the virus in vague terms. "It is related to Marburg. It is transmitted the same way, through contact with infected tissue and blood, and the signs and symptoms are much the same."

"How bad is it?"

"The case-fatality rate is fifty to ninety percent."

Dalgard understood exactly what that meant. The virus was much worse than Marburg.

C. J. continued, "With the information we have, we are going to notify state and national public health officials."

Dalgard spoke carefully. "Would you, *ahem,* would you please wait until seven p.m., to allow me to apprise my corporate headquarters of recent developments?"

C. J. agreed to wait before pulling the trigger, though in fact General Russell had already called the C.D.C. Now C. J. had a favor to ask of Dalgard. Would it be all right if he sent someone down to Reston tomorrow to have a look at some samples of dead monkeys?

Dalgard resisted. He had sent a little bit of blood and tissue to the Army for diagnosis—and look what was happening. This thing could go way out of control. He sensed that Colonel Peters was not telling him all there was to know about this virus called Ebola. Dalgard feared he could lose control of the situation in a hurry if he let the Army get its foot in the door. ''Why don't we meet by phone early tomorrow morning and discuss this approach?'' Dalgard replied.

After the phone call, C. J. Peters found Nancy Jaax and asked her if she would come with him to meet Dalgard the next day and look at some monkey tissue. He assumed Dalgard would give permission. She agreed to go with him.

Nancy Jaax walked across the parade ground back to the Institute and found Jerry in his office. He looked up at her with a pained expression on his face. He had been staring out the window and thinking about his murdered brother. It was dark; there was nothing to see out there except a blank wall.

She closed the door. ''I've got something for you. This is close hold. This is hush-hush. You are not going to believe this. There's Ebola virus in a monkey colony in Virginia.''

They drove home, talking about it, traveling north on the road that led to Thurmont along the foot of Catoctin Mountain.

''This is killing me—I'll never get away from this bug,'' she said to him.

It seemed clear that they both were going to be involved in the Army action. It wasn't clear what kind of an action it would be, but certainly something big was going to go down. She told Jerry that tomorrow she would probably visit the monkey house with C. J. and that she would be looking at monkey tissues for signs of Ebola.

Jerry was profoundly surprised: so this was what Nancy's work with Ebola had come to. He was impressed with his wife and bemused by the situation. If he was worried about her, he didn't show it.

They turned up a gentle swing of road that ran along the

side of the mountain, and passed through apple orchards, and turned into their driveway. It was eight o'clock, and Jason was home. Jaime had gone off to her gymnastics practice. The kids were latchkey children now.

Jason was doing his homework. He had made himself a microwave dinner of God knows what. Their son was a self-starter, a little bit of a loner, and very self-sufficient. All he needed was food and money, and he ran by himself.

The two colonels changed out of their uniforms into sweat clothes, and Nancy put a frozen chunk of her homemade stew into the microwave and thawed it. When the stew was warm, she poured it into a Thermos jar. She put the dog and the Thermos into the car, and she drove out to get Jaime at her gymnastics practice. The gym was a half hour's drive from Thurmont. Nancy picked up Jaime and gave her the stew to eat in the car. Jaime was an athletic girl, short, dark haired, sometimes inclined to worry about things—and she was exhausted from her workout. She ate the stew and fell asleep on the back seat while Nancy drove her home.

The Colonel Jaaxes had a water bed, where they spent a lot of time. Jaime got into her pajamas and curled up on the water bed next to Nancy and fell asleep again.

Nancy and Jerry read books in bed for a while. The bedroom had red wallpaper and a balcony that overlooked the town. They talked about the monkey house, and then Nancy picked up Jaime and carried her into her own bedroom and tucked her into her bed. Around midnight, Nancy fell asleep.

Jerry continued to read. He liked to read military history. Some of the most brutal combat in history had occurred in the rolling country around Catoctin Mountain: at the cornfield at Antietam, where every individual stalk of corn had been slashed away by bullets, and where the bodies had lain so thick a person could walk on them from one end of the cornfield to the other. He could look out his bedroom window and imagine the blue and gray armies crawling across the land. That night he happened to be reading *The Killer Angels,* a novel by Michael Shaara about the Battle of Gettysburg:

> Then Lee said slowly, "Soldiering has one great trap."
> Longstreet turned to see his face. Lee was riding slowly ahead, without expression. He spoke in that same slow voice.
> "To be a good soldier you must love the army. But to be a good officer you must be willing to order the death of the thing you love. That is . . . a very hard thing to do. No other profession requires it. That is one reason why there are so few very good officers. Although there are many good men."

He switched out the light, but he could not sleep. He rolled over, and the water bed gurgled. Every time he closed his eyes, he thought about his brother, John, and he saw in his mind's eye an office splattered with blood. Eventually it was two o'clock in the morning, and he was still awake, thinking to himself, I'm just laying here in the dark, and nothing's happening.

Garbage Bags

NOVEMBER 29, WEDNESDAY

Dan Dalgard slept peacefully that night, as he always did. He had never heard of Ebola virus, but the brief conversation with Colonel C. J. Peters had given him the basic picture. He had been around monkeys and monkey diseases for a long time, and he was not particularly frightened. Many days had passed during which he had been exposed to infected blood, and he certainly had not become sick yet.

Early in the morning, his telephone rang at home. It was Colonel Peters calling. Again Peters asked him if he could send some people down to look at specimens of tissue from the monkeys. Dalgard said that would be all right. Peters then repeated his request to see the monkey house. Dalgard turned away the question and wouldn't answer it. He didn't know Peters, and he wasn't going to open any doors to him until he had met the man and had a chance to size him up.

He drove down Leesburg Pike to work, turned through a gate, parked his car, and went into the main building of Hazleton Washington. His office was a tiny cubicle with a glass wall that looked across the lawn; his door looked back to a secretarial pool, a cramped area where you could hardly move around without bumping into people. There was no privacy in Dalgard's office; it was a fishbowl. He tended to

spend a lot of time looking out the window. Today he behaved with deliberate calm. No one in the office detected any
unusual emotion, any fear.

He called Bill Volt, the manager of the monkey house.
Volt gave him a shocking piece of news. One of the animal
caretakers was very sick, might be dying. During the night,
the man had had a heart attack and had been taken to
Loudon Hospital, not far away. There's no further information, Volt said, and we're still trying to find out what happened. He's in the cardiac-care unit, and no one can talk to
him. (The man's name will be given here as Jarvis Purdy. He
was one of four workers in the monkey house, not including
Volt.)

Dalgard was extremely dismayed and couldn't rule out the
possibility that the man was breaking with Ebola. A heart
attack is usually caused by a blood clot in the heart muscle.
Had he thrown a clot that had lodged in his heart? Could
Ebola cause you to throw a clot? Was Jarvis Purdy clotting
up? Suddenly Dalgard felt as if he was losing control of the
situation.

He told Bill Volt that he was to suspend all unnecessary
activity in the monkey rooms. As he later recorded in his
diary:

> All operations other than feeding, observation and cleaning were to
> be suspended. Anyone entering the rooms was to have full protec-
> tion—Tyvek suit, respirator, and gloves. Dead animals were to be
> double-bagged and placed in a refrigerator.

He also mentioned to Volt that the news media were almost
certainly going to get onto this story. He told Volt that he
didn't want any employees to go outside the building wearing their biohazard gear. If pictures of Hazleton workers
wearing face masks and white suits wound up on the evening
news, it could cause a panic.

Dalgard called the hospital and reached Purdy's doctor.
The doctor said that Purdy's condition was guarded but sta-

ble. Dalgard told the doctor that if any aspect of Purdy's heart attack wasn't typical, he should please call Colonel C. J. Peters at Fort Detrick. He was careful not to mention the word *Ebola.*

Later that morning, C. J. Peters and Nancy Jaax headed out from Fort Detrick for Virginia, and Gene Johnson came with them. The officers wore their uniforms, but they drove in civilian cars so as not to attract attention. The traffic moved slowly. It was a clear, cold, windy day. The grass along the road was wet and green, still growing, untouched by frost. They turned off Leesburg Pike at the Hazleton offices. Dalgard met them in the lobby and escorted them to another building, which was a laboratory. There a pathologist had prepared a set of slides for Nancy to look at. The slides contained slices of liver from monkeys that had died in the monkey house.

She sat down at a microscope, adjusted the eyepieces, and began to explore the terrain. She zoomed around and paused. The terrain was a mess. Something had ravaged these cells. They were blitzed and pock-marked, as if the liver had been carpet bombed. Then she saw the dark blobs in the cells— the shadows that did not belong there. They were crystalloids. And they were huge.

This was extreme amplification.

"Oh, fuck," she said in a low voice.

The bricks did not look like crystals. Ebola bricks come in all kinds of shapes—horseshoes, blobs, lumps, even rings. Some of the cells consisted of a single brick, a huge mother of a brick, a brick that had grown so fat that the whole cell had plumped up. She saw clusters of cells packed with bricks. She saw rotten pockets where all the cells had popped and died, forming a liquefied spot that was packed with wall-to-wall bricks.

While she looked at the slides, C. J. Peters and Gene Johnson took Dan Dalgard aside and questioned him closely about the use of needles at the monkey house. The Ebola virus had spread in Zaire through dirty needles. Had the company been giving monkeys shots with dirty needles?

Dalgard was not sure. The company had an official policy of always using clean needles. "Our policy is to change needles after every injection," he said. "Whether it is done religiously is anybody's guess."

Nancy collected some pieces of sterilized liver and spleen that were embedded in wax blocks, and she put the blocks in a Styrofoam cup to take back to Fort Detrick for analysis. These samples were exceedingly valuable to her and to the Army. What would be even more valuable would be a sample containing live virus.

C. J. Peters asked Dalgard again if they could all go see the monkey house.

"Well—let's not go there now," Dalgard replied. He made it clear to the officers that the building was private property.

"What about some samples of monkey? Can we get some samples?" they asked.

"Sure," Dalgard said. He told them to drive out Leesburg Pike in the direction of the monkey house. There was an Amoco gas station on the pike, he said, and the colonels were to park their cars there and wait. "A guy is going to come and meet you. He'll bring some samples with him. And he can answer your questions," he said.

"The samples ought to be wrapped in plastic and put in boxes for safety," C. J. said to Dalgard. "I want you to do that."

Dalgard agreed to wrap the samples in plastic.

Then C. J., Nancy, and Gene drove out to the gas station, where they parked in a cul-de-sac by the highway, near some pay telephones. By now it was early afternoon, and they were hungry—they had missed lunch. Nancy went into the gas station and bought Diet Cokes for everyone and a pack of cheddar-cheese crackers for herself, and she bought C. J.

some peanut-butter crackers. The Army people sat in their
two cars, eating junk food, feeling cold, and hoping that
someone would show up soon with samples of monkey.

C. J. Peters observed the comings and goings at the gas
station. It gave him a sense of life and time passing, and he
enjoyed the pleasant normality of the scene. Truckers
stopped for diesel and Cokes, and businesspeople stopped
for cigarettes. He noticed an attractive woman park her car
and go over to one of the pay telephones, where she spoke at
length to someone. He whiled away the time imagining that
she was a housewife talking to a boyfriend. What would
these people think if they knew what had invaded their town?
He had begun to think that the Army might have to act
decisively to put out this fire. He had been in Bolivia when a
hot agent called Machupo had broken out, and he had seen a
young woman die, covered with blood. North America had
not yet seen an emergence of an agent that turned people
into bleeders. North America was not ready for that, not yet.
But the possibilities for a huge break of Ebola around Washington were impressive when you thought about it.

He wondered about AIDS. What would have happened if
someone had noticed AIDS when it first began to spread? It
had appeared without warning, secretly, and by the time we
noticed it, it was too late. If only we had had the right kind of
research station in central Africa during the nineteen-seventies . . . we might have seen it hatching from the forest. If
only we had seen it coming . . . we might have been able
to stop it, or at least slow it down; . . . we might have been
able to save at least a hundred million lives. At least. Because the AIDS virus's penetration of the human species was
still in its early stages, and the penetration was happening
inexorably. People didn't realize that the AIDS thing had only
just begun. No one could predict how many people were
going to die of AIDS, but he believed that the death toll, in the
end, could hit hundreds of millions—and that possibility had
not sunk in with the general public. On the other hand, sup-

pose AIDS had been noticed? Any "realistic" review of the AIDS virus when it was first appearing in Africa would probably have led experts and government officials to conclude that the virus was of little significance for human health and that scarce research funds should not be allocated to it—after all, it was just a virus that infected a handful of Africans, and all it did was suppress their immune systems. So what? And then the agent had gone on a tremendous amplification all over the planet, and it was still expanding its burn, with no end in sight.

We didn't really know what Ebola virus could do. We didn't know if the agent in the monkey house was, in fact, Ebola Zaire or if it was something else, some new strain of Ebola. An agent that could travel in a cough? Probably not, but who could tell? The more he thought about it, the more he wondered, Who is going to take out those monkeys? Because someone is going to have to go in there and take them out. We can't just walk away from that building and let it self-destruct. This is a human-lethal virus. Who is going to sack the monkeys? The guys who work for the company?

He had begun to wonder whether the Army should move in with a military biohazard SWAT team. His own term for this type of action was *nuke*. To nuke a place means to sterilize it, to render it lifeless. If the hosts are people, you evacuate them and put them in the Slammer. If the hosts are animals, you kill them and incinerate the carcasses. Then you drench the place with chemicals and fumes. He wondered if the Army would have to nuke the monkey house.

Gene Johnson sat in the passenger seat next to C. J. Peters. His mind was somewhere else. His mind was in Africa. He was thinking about Kitum Cave.

Gene was very worried about this situation, not to say shit scared. He thought to himself, I don't know how we are going to get out of this one without people dying. His worry was growing all the time, every minute. The U.S. military, he thought, is stepping into a crisis that is already full-blown, and if something goes wrong and people die, the military will be blamed.

Suddenly he turned to C. J. and spoke his mind. He said, "It looks inevitable that we're going to have to take out all the monkeys. A Level 4 outbreak is not a game. I want to warn you about just how detailed and major an effort this is going to be. It's going to be very complex, it's going to take some time, and we have to be very fucking careful to do it right. If we are going to do it right, the gist of what I'm saying, C. J., is that we can*not* have amateurs in key positions. We need to have experienced people who know what they are doing. Do you understand what's going to happen if something goes wrong?'' And he was thinking: Peters— Peters—he's never been in an outbreak this complicated— none of us has—the only thing like it was Kitum Cave. And Peters wasn't there.

C. J. Peters listened to Gene Johnson in silence, and didn't reply. He felt that it was sort of irritating to get this kind of advice from Gene—when he's telling you the obvious, telling you what you already know.

C. J. Peters and Gene Johnson had a stressful, complicated relationship. They had journeyed together in a truck expedition across central Africa, looking for Ebola virus, and a lot of tension had built up between the two men by the end of the trip. The traveling had been brutal, as hard as any on earth—roads didn't exist, bridges were gone, the maps must have been drawn by a blind monk, the people spoke languages not even the native translators could understand, and the expedition had not been able to find enough food and water. Worst of all, they ran into difficulty finding human cases of Ebola—they were not able to discover the virus in a natural host or in people.

It was during that trip, perhaps as a result of the chronic food shortage, that C. J. had taken to eating termites. The ones that swarmed out of their nests. They had wings. Gene, who was more fastidious than C. J., had not been quite so eager to try them. Popping termites in his mouth, C. J. would make remarks like, "They have this extra . . . *mmm* . . . ," and he would smack his lips, *smack, smack,* and you'd hear a mouthful of termites crunching between his teeth, and he'd spit out the wings, *pah, ptah*. The African members of the expedition, who liked termites, had pushed Gene to try them, too, and finally he did. He placed a handful of them in his mouth, and was surprised to find that they tasted like walnuts. C. J. had spoken longingly of finding the African termite queen, the glistening white sac that was half a foot long and as thick as a bratwurst, bursting with eggs and creamy insect fat, the queen you ate alive and whole, and she was said to twitch as she went down your throat. Although snacking on termites had amused them, they had argued with each other about how to do the science, how to search for the virus. In Africa, Gene had felt that C. J. was trying to run the show, and it irritated Gene to no end.

Suddenly a blue, windowless, unmarked van turned off the road and pulled through the gas station and parked next to them. The van parked in such a way that no one on the road or at the gas station could see what went on between the two vehicles. A man swung heavily out of the driver's seat. It was Bill Volt. He walked over to the Army people, and they got out of their cars.

"I've got 'em right back here," he said, and he threw open the side door of the van.

They saw seven black plastic garbage bags sitting on the floor of the van. They could see the outlines of limbs and heads in the bags.

C. J. said to himself, What *is* this?

Nancy gritted her teeth and silently pulled in a breath. She could see how the bags bulged in places, as if liquid had pooled inside them. She hoped it wasn't blood. "What on earth is all of that?" she exclaimed.

"They died last night," Volt said. "They're in double bags."

Nancy was getting a nasty feeling in the pit of her stomach. "Has anybody cut himself fooling around with these monkeys?" she asked.

"No," Volt replied.

Then Nancy noticed that C. J. was looking sideways at her. It was a significant look. The message was, So who's going to drive the dead monkeys back to Fort Detrick?

Nancy stared back at C. J. He was pushing her, and she knew it. They were both division chiefs at the Institute. He outranked her, but he was not her boss. He can push me just so far, and I can push him right back. "I'm not putting that shit in the trunk of my car, C. J.," she said. "As a veterinarian, I have certain responsibilities with regard to the transportation of dead animals, sir. I can't just knowingly ship a dead animal with an infectious disease across state lines."

Dead silence. A grin spread over C. J.'s face.

"I agree that it needs to be done," Nancy went on. "You're a doc. You can get away with this." She nodded at his shoulder boards. "This is why you put on those big eagles."

They burst into nervous laughter.

C. J. inspected the bags—it was a relief to see that the monkeys were double-bagged or triple-bagged—and he decided to take them back to Fort Detrick and worry about health laws afterward. His reasoning, as he explained to me later, went like this: "If the guy drove them back to the Reston monkey facility, I felt there would be a certain added risk to the population just from his driving them around in the van, and there would be a delay in diagnosing them. We felt that if we could quickly get a definite diagnosis of Ebola it would be in everyone's favor." Surely some smart Army lawyers could figure out why the act of carrying Ebola-ridden dead monkeys across state lines in the trunk of a private automobile was so completely legal that there had never even been any question about it.

His old red Toyota was not in the best of shape, and he had lost any interest in its resale value. He popped the trunk. It was lined with carpet, and he didn't see any sharp edges anywhere that might puncture a plastic bag.

They didn't have rubber gloves. So they would do the lifting bare-handed. Nancy, keeping her face well away from the enclosed air of the van, inspected the outsides of the bags for any droplets of blood. "Have the exteriors of the bags been disinfected?" she asked Volt.

Volt said he'd washed the outsides of the bags with Clorox bleach.

She held her breath, fighting the puke factor, and picked up a bag. The monkey kind of slid around inside it. They piled the bags one by one gently in the Toyota's trunk. Each monkey weighed between five and twelve pounds. The total weight came to around fifty pounds of Biohazard Level 4 liquefying primate. It depressed the rear end of the Toyota. C. J. closed the trunk.

Nancy was anxious to dissect the monkeys right away. If you left an Ebola monkey inside a plastic bag for a day, you'd end up with a bag of soup.

"Follow behind me, and watch for drips," C. J. joked.

Space Walk

They arrived at the Institute in midafternoon. C. J. Peters parked beside a loading dock on the side of the building and found some soldiers to help him carry the garbage bags to a supply air lock that led to the Ebola suite. Nancy went to the office of a member of her staff, a lieutenant colonel named Ron Trotter, and told him to suit up and go in; she would follow. They would be buddies in the hot zone.

As she always did before going into Level 4, she took off her engagement ring and her wedding band, and locked them away in her desk. She and Trotter walked down the hall together, and he went first into the small locker room that led to AA-5 while she waited in the corridor. A light went on, telling her that he had gone on to the next level, and she swiped her security card across a sensor, which opened the door into the locker room. She took off all of her clothes, put on a long-sleeved scrub suit, and stood before the door that led inward, blue light falling on her face. Beside the door there was another security sensor. This one was a numerical key pad. You can't bring your security card with you into the higher levels. A security card would be melted or ruined by chemicals during the decontamination process. Therefore you memorize your security code. She punched a string of numbers on the key pad, and the building's central computer noticed that Jaax,Nancy, was attempting entry. Finding that she was cleared to enter AA-5, the computer unlocked the door and beeped to let her know that she could proceed inward without setting off alarms. She walked through the shower stall into the bathroom, put on white socks, and continued inward, opening a door that led to the Level 3 staging area.

There she met Lieutenant Colonel Trotter, a stocky, dark-haired man whom Nancy had worked with for many years. They put on their inner gloves and taped their cuffs. Nancy put a pair of hearing protectors over her ears. She had started wearing them a while back, when people had begun to suspect that the roar of air in your suit might be loud enough to damage your hearing. They hauled on their space suits and sealed the Ziploc zippers. They edged around each other as they fiddled with their suits. People wearing biohazard space suits tend to step around one another like two wrestlers at the beginning of a match, watching the other person's every move, especially watching the hands to make sure they don't hold a sharp object. This cringing becomes instinctive.

They closed up their suits and lumbered across the staging area to a large air-lock door. This was a supply air lock. It did not lead into the hot zone. It led to the outside world. They opened it. On the floor of the air lock sat the seven garbage bags.

"TAKE AS MANY AS YOU CAN CARRY," she said to Lieutenant Colonel Trotter.

He picked up a few bags, and so did she. They shuffled back across the staging area to the air-lock door that led to Level 4. She picked up a metal pan containing tools. She was getting warm, and her faceplate fogged up. They opened the air-lock door and stepped in together. Nancy took a breath and gathered her thoughts. She imagined that passing through the gray-zone door into Level 4 was like a space walk, except that instead of going into outer space, you went into inner space, which was full of the pressure of life trying to get inside your suit. People went into Level 4 areas all the time at the Institute, particularly the civilian animal caretakers. But going into a containment zone to perform a necropsy on an animal that had died of an amplified unknown hot agent was something a little different. This was high-hazard work.

Nancy centered herself and brought her breathing under control. She opened the far door and went through to the hot side. Then she reached back inside the air lock and pulled the chain in the chemical shower. That started a decon cycle running in the air lock that would eliminate any hot agents that might have leaked into the air lock as they were going through.

They put on their boots and headed down the cinder-block hallway, lugging the monkeys. Their air was going stale inside their space suits, and they needed to plug in right away.

They came to a refrigerator room, and put all the bags in the refrigerator except for one. This bag they carried into the necropsy room. Stepping around each other cautiously, they plugged in their air hoses, and dry air cleared their faceplates. The air thundered distantly beyond Nancy's hearing protectors. They gloved up, pulling surgical gloves over their space-suit gloves. She laid her tools and specimen containers at the head of the table, counting them off one by one.

Trotter untwisted some ties on the garbage bag and opened it, and the hot zone inside the bag merged with the hot zone of the room. He and Nancy together lifted the monkey out and laid it on the dissection table. She switched on a surgical lamp.

Unclouded brown eyes stared at her. The eyes looked normal. They were not red. The whites were white, and the pupils were clear and black, dark as night. She could see a reflection of the lamp in the pupils. Inside the eyes, behind the eyes, there was nothing. No mind, no existence. The cells had stopped working.

Once the cells in a biological machine stop working, it can never be started again. It goes into a cascade of decay, falling toward disorder and randomness. Except in the case of viruses. They can turn off and go dead. Then, if they come in contact with a living system, they switch on and multiply. The only thing that "lived" inside this monkey was the unknown agent, and it was dead, for the time being. It was not multiplying or doing anything, since the monkey's cells were dead. But if the agent touched living cells, Nancy's cells, it would come alive and begin to amplify itself. In theory, it could amplify itself around the world in the human species.

She took up a scalpel and slit the monkey's abdomen, making a slow and gentle cut, keeping the blade well away from her gloved fingers. The spleen was puffed up and tough, leathery, like a globe of smoked salami. She did not see any bloody lesions inside this monkey. She had expected that the monkey's interior would be a lake of blood, but no, this monkey looked all right, it had not bled into itself. If the animal had died of Ebola, this was not a clear case. She opened up the intestine. There was no blood inside it. The gut looked okay. Then she examined the stomach. There she found a ring of bleeding spots at the junction between the stomach and the small intestine. This could be a sign of Ebola, but it was not a clear sign. It could also be a sign of simian fever, not Ebola. Therefore, she could not confirm the presence of Ebola virus in this animal based on a visual inspection of the internal organs during necropsy.

Using a pair of blunt scissors, she clipped wedges out of the liver and pressed them on glass slides. Slides and blood tubes were the only glass objects allowed in a hot zone, because of the danger of glass splinters if something broke. All laboratory beakers in the room were made of plastic.

She worked slowly, keeping her hands out of the body cavity, away from blood as much as possible, rinsing her gloves again and again in a pan of EnviroChem. She changed her gloves frequently.

Trotter glanced at her once in a while. He held the body open for her and clamped blood vessels, handing her tools when she asked for them. They could read each other's lips.

"FORCEPS," she mouthed silently, pointing to it. He nodded and handed her a forceps. They did not talk. She was alone with the sound of her air.

She was beginning to think that this monkey did not have Ebola virus. In biology, nothing is clear, everything is too complicated, everything is a mess, and just when you think you understand something, you peel off a layer and find deeper complications beneath. Nature is anything but simple. This emerging virus was like a bat crossing the sky at evening. Just when you thought you saw it flicker through your field of view, it was gone.

Shoot-out

While Nancy Jaax was working on the monkeys, C. J. Peters was in the conference room at Fort Detrick's headquarters building. Careers were at stake in this room. Almost all of the people in the world who understood the meaning of Ebola virus were sitting around a long table. General Russell sat at the head of the table, a tall, tough-looking figure in uniform; he chaired the meeting. He did not want the meeting to turn into a power struggle between the Centers for Disease Control and the Army. He also did not want to let the C.D.C. take over this thing.

Dan Dalgard was there, wearing a dark suit, seeming reserved and cool; in fact, he churned with nervousness. Gene Johnson glowered over the table, bearded and silent. There were officials from the Virginia Department of Health and from Fairfax County. Fred Murphy—the codiscoverer of Ebola virus, the C.D.C. official whom General Russell had called—sat at the table beside another official from the C.D.C., Dr. Joseph B. McCormick.

Joe McCormick was the chief of the Special Pathogens Branch of the C.D.C., the branch that had been run by Karl Johnson, another codiscoverer of Ebola virus. Joe McCormick was the successor to Karl Johnson—he had been appointed to the job when Johnson retired. He had lived and worked in Africa. He was a handsome, sophisticated medical doctor with curly dark hair and round Fiorucci spectacles, a brilliant, ambitious man, charming and persuasive, with a quick, flaring temper, who had done extraordinary things in his career. He had published major research articles on Ebola. Unlike anyone else in the room, he had seen and treated human cases of Ebola virus.

It happened that Joe McCormick and C. J. Peters couldn't stand each other. There was bad blood between these two doctors that went back many years. They had both rifled the darkest corners of Africa searching for Ebola, and neither of them had found its natural hiding place. Like Peters, Joe McCormick evidently felt that now, finally, he was closing in on the virus and getting ready to make a spectacular kill.

The meeting began with Peter Jahrling, the codiscoverer of the strain that burned in the monkeys. Jahrling stood up and spoke, using charts and photographs. Then he sat down.

Now it was Dalgard's turn to speak. He was exceedingly nervous. He described the clinical signs of disease that he had seen at the monkey house, and by the end he felt that no one had noticed his nervousness.

Immediately afterward, Joe McCormick got up and spoke. What he said remains a matter of controversy. There is an Army version and there is another version. According to Army people, he turned to Peter Jahrling and said words to this effect: Thanks very much, Peter. Thanks for alerting us. The big boys are here now. You can just turn this thing over to us before you hurt yourselves. We've got excellent containment facilities in Atlanta. We'll just take all your materials and your samples of virus. We'll take care of it from here.

In other words, the Army people thought McCormick

tried to present himself as the only real expert on Ebola. They thought he tried to take over the management of the outbreak and grab the Army's samples of virus.

C. J. Peters fumed, listening to McCormick. He heard the speech with a growing sense of outrage, and thought it was "very arrogant and insulting."

McCormick remembers something different. "I'm sure I offered some help or assistance with the animal situation at Reston," he recalled, when I telephoned him. "I don't know that there was any conflict. If there was any animosity, it came from *their* side, not ours, for reasons they know better than I. Our attitude was, Hey guys, good work."

But McCormick and the Army had not been getting along well, and there was a history of conflict. In the past, McCormick had publicly criticized Gene Johnson, the Army's Ebola expert, for spending a lot of money to explore Kitum Cave and then not publishing his findings. McCormick expressed his opinion of the Army to me this way: "They want to tell you about their experiments, but the way to tell people about them is to *publish* them. That's not an unreasonable criticism. They're spending taxpayers' money." And besides, "None of them had spent as much time in the field as I had. I was one of those who had dealt with human cases of Ebola. No one else there had done that."

What McCormick had done was this. In 1979, reports reached the C.D.C. that Ebola had come out of hiding and was burning once again in southern Sudan, in the same places where it had first appeared, in 1976. The situation was dangerous, not only because of the virus but because a civil war was going on in Sudan—the area where Ebola raged also happened to be a war zone. McCormick nevertheless volunteered to go there to try to collect some human blood and bring the strain back alive to Atlanta. He traveled to Sudan in the company of another C.D.C. doctor named Roy Baron. McCormick and Baron arrived in southern Sudan in a light plane flown by two terrified bush pilots. Around sunset, they landed at an airstrip near a Zande village. The pilots were too scared to get out of the plane. It was getting dark, and the pilots decided to spend the night in the cockpit, sitting on the airstrip. They warned McCormick and Baron that they would leave the next morning at sunrise. The doctors had until dawn to find the virus.

They shouldered their backpacks and walked into the village, looking for Ebola. They arrived at a mud hut. Villagers stood around the hut, but wouldn't go inside. They heard sounds of human agony. A dark doorway led inside. They couldn't see into the hut, but they knew that Ebola was in there. McCormick rummaged in his backpack and found his flashlight, but it was dead, and he realized that he had forgotten to bring batteries. He asked the crowd if anyone had a light, someone brought him a lantern, and they entered the hut.

Years later, McCormick told me that he would never forget the sight. The first thing he saw was a number of red eyes staring at him. The air inside the hut reeked of blood. People lay on straw mats on the floor. Some were having convulsions—the final phase, as death sets in—their bodies rigid and jerking, their eyes rolled up into the head, blood streaming out of the nose and flooding from the rectum. Others had gone into terminal comas, and were motionless and bleeding out. The hut was a hot zone.

McCormick opened his backpack and fished out rubber gloves, a paper surgical gown, a paper surgical mask, and paper boots to cover his shoes, to keep them from becoming wet with blood. After he had dressed himself, he laid out his blood tubes and syringes on a mat. Then he began drawing blood from people. He worked all night in the hut on his knees, collecting blood samples and taking care of the patients as best he could. Baron worked at his side.

Sometime during the night, McCormick was drawing blood from an old woman. Suddenly she jerked and thrashed, having a seizure. Her arm lashed around, and the bloody needle came out of her arm and jabbed into his thumb. Uh-oh, he thought. That would be enough to do it. The agent had entered his bloodstream.

At dawn, they gathered up their tubes of blood and ran to the airplane and handed the samples to the pilots. The question for McCormick was what to do with himself, now that he had been pricked with a bloody needle. That was a massive exposure. He probably had three to four days before he broke with Ebola. Should he leave Sudan now, get himself to a hospital? He had to make a decision—whether to leave with the pilots or stay with the virus. It seemed obvious that the pilots would not come back later to pick him up. If he planned to leave and get medical help for himself, the time to do it was now. There was an additional factor. He was a physician, and those people in the hut were his patients.

He returned to the village with Baron, and rested that day in a hut. That evening, he and his colleague had dinner with some local United Nations officials, where McCormick drank at least half a bottle of scotch. He got talkative, then he collapsed. Baron dragged and carried McCormick to a nearby hut, sat him up on a cot, and gave him a large transfusion of blood serum from Africans who had survived Ebola. This might help McCormick fight off the virus. Or it might not. That night, whirling drunk on scotch, McCormick still could not sleep. He lay awake, thinking about the needle jabbing into his thumb, thinking about Ebola starting its inevitable replication in his bloodstream.

He worked with Ebola patients for the next four days inside the hut, and still he did not have a headache. Meanwhile, he watched the old lady like a hawk to see what happened to her. On the fourth day, to his surprise, the old lady recovered. She had not had Ebola. She had probably been suffering from malaria. She had not been having an Ebola seizure but, rather, had been shivering from a fever. He had walked away from a firing squad.

Now, at the meeting at Fort Detrick, Joe McCormick of

SHOOT-OUT   201

the C.D.C. was convinced that Ebola virus does not travel easily, especially not through the air. He had not become sick, even though he had breathed the air inside an Ebola-ridden hut for days and nights on end. He felt strongly that Ebola is a disease that is not easy to catch. Therefore, in his view, it was not as dangerous as perhaps the Army people believed.

Dan Dalgard asked a question of the assembled experts. He said, "How soon after we give you samples can you tell us whether they have virus in them?"

C. J. Peters replied, "It may take a week. This is all we know."

Joe McCormick spoke up. Wait a minute, he said—he had a new, fast probe test for Ebola virus that would work in just twelve hours. He argued that the C.D.C. should have the virus and the samples.

C. J. Peters turned and stared at McCormick. C. J. was furious. He didn't believe McCormick had any quick test for Ebola. He thought it was Joe McCormick blowing smoke, trying to get his hands on the virus. He thought it was a poker bluff in a high-stakes game for control of the virus. It was a delicate situation, because how could he say in front of all these state health officials, "Joe, I just don't believe you"? He raised his voice and said, "An ongoing epidemic is not the time to try to field-test a new technique." He argued that Fort Detrick was closer to the outbreak than was the C.D.C., in Atlanta, and therefore it was appropriate for the Army to have the samples and try to isolate the virus. What he did not say—no reason to rub it in—was that seven dead monkeys were at that very moment being examined by Nancy Jaax. Even as they argued, she was exploring the monkeys. What's more, the Army was growing the virus in cultures. Possession is nine tenths of the law, and the Army had the meat and the agent.

Fred Murphy, the other C.D.C. man, was sitting next to McCormick. He began to realize that the C.D.C. was not in a good position to argue the matter. He leaned over and whispered, "Joe! Calm your jets. Stifle it, Joe. We're outnumbered here."

General Philip Russell had been sitting back, watching the argument, saying nothing. Now he stepped in. In a calm but almost deafeningly loud voice he suggested that they work out a compromise. He suggested that they split the management of the outbreak.

A compromise seemed to be the best solution. The general and Fred Murphy quickly worked out the deal, while McCormick and Peters stared at each other with little to say. It was agreed that the C.D.C. would manage the human-health aspects of the outbreak and would direct the care of any human patients. The Army would handle the monkeys and the monkey house, which was the nest of the outbreak.

The Mission

Colonel C. J. Peters now felt that he had permission to get the action under way. As soon as the meeting broke up, he began to line up his ducks. The first thing he needed was a field officer who could lead a team of soldiers and civilians into the monkey house. He needed to form a military-action unit.

He had already decided who was going to lead the mission. It was going to be Colonel Jerry Jaax, Nancy's husband. Jerry had never worn a space suit, but he was the chief of the veterinary division at the Institute, and he understood monkeys. His people, both soldiers and civilians, were certainly going to be needed. No one else had the training to handle monkeys.

He found Jerry in his office, staring out the window and chewing on a rubber band. C. J. said, "Jerry, I believe we have a situation down in Reston." A *situation*. Code for a hot agent. "It looks like we're going to have to go down and take those monkeys out, and we're going to do it in Biosafety Level 4 conditions." He asked Jerry to assemble teams of soldiers and civilian employees to be ready to move out with space suits in twenty-four hours.

Jerry walked over to Gene Johnson's office and told him

that he'd been put in charge of the mission. The office was a mess. He wondered how Gene, as large a man as he was, could even fit himself in among the stacks of paper.

Jerry and Gene immediately began to plan a biohazard operation. There had been a general decision to take out one room of monkeys, and see how that worked, see how things went—see if the virus was spreading. They set up their priorities.

Priority One. Safety of the human population.

Priority Two. Euthanasia of the animals with a minimum of suffering.

Priority Three. Gathering of scientific samples. Purpose: to identify the strain and determine how it travels.

Gene felt that if the team did its job properly, the human population of Washington would be safe. He put on his glasses and hunched over and fished through his papers, his beard crushed on his chest. He knew already that he was not going to go inside that building. No way in hell. He had seen monkeys die too many times, and he could not bear it anymore. In any case, his job was to gather equipment and people and move them into the building, and then to extract the people and equipment and dead animals safely.

He had saved lists, long lists of all the gear he had brought to Kitum Cave. He pawed through his papers, swearing gently. He had literally tons of African gear. He had squirreled it away in all kinds of hiding places at the Institute, where other people couldn't find it and rip it off.

Gene was terribly excited, and also afraid. His nightmares about Ebola virus, the bad dreams of liquid running through pinholes into his space suit, had never really gone away. He would still wake up at night thinking, My God, there's been an exposure. He had spent almost ten years hunting Ebola and Marburg in Africa, with little success, and suddenly one of the bastards had reared its head in Washington. His favorite saying came back to him: "Chance favors the prepared mind." Well, the chance had come. If a piece of gear had been handy in Kitum Cave, it would be handy in the monkey house. As Gene thought about it, he realized that the building was very much like Kitum Cave. It was an enclosed air space. Dead air. Air-handling system broken, failed. Dung all over the place. Monkey urine in pools. A hot cave near Washington. And there were people who had been inside the cave who might be infected with the virus by now. How would you move your teams in and out of the hot area? You would have to set up a staging area. You would have to have a gray area—an air lock with a chemical shower of some kind. Somewhere inside that building lived a Level 4 life form, and it was growing, multiplying, *cooking* inside hosts. The hosts were monkeys and, perhaps, people.

2000 HOURS, WEDNESDAY

Dan Dalgard left USAMRIID and drove back to his office on Leesburg Pike, arriving there around eight o'clock. The office was deserted; everyone had gone home. He straightened up his desk, shut down his computer, and removed a floppy disk that contained his daily diary, his "Chronology of Events." He put the disk into his brief case. He said good night to a security guard at the front desk and drove home. On the road, he realized that he had forgotten to call his wife to tell her that he would be late. He stopped at a Giant Food supermarket and bought her a bunch of cut flowers, carnations and mums. When he arrived home, he reheated his dinner in the microwave and joined his wife in the family room, where he ate sitting in a recliner chair. He was exhausted. He put another log into the wood stove and sat down at his personal computer, which was located next to his clock-repair bench. He inserted the floppy disk and began typing. He was bringing his diary up to date.

So much had happened during the day that he had difficulty keeping it all straight in his mind. In the morning, he had learned that the monkey caretaker named Jarvis Purdy was in the hospital, reportedly with a heart attack. Jarvis was resting comfortably, and there had been no reports that his condition was getting worse. Should I have notified the hospital that Jarvis might be infected with Ebola? If he does have Ebola, and it spreads within the hospital, am I liable? Jesus! I'd better get someone to go over to the hospital first thing tomorrow and tell Jarvis what's going on. If he hears it on the news first, he's liable to have another heart attack!

He had gotten all the other monkey caretakers fitted with respirators, and he had briefed them on what was known about the transmission of Ebola and Marburg to humans, and he had suspended all daily operations in the building other than feedings once a day, observation, and cleaning of the animal rooms. He had briefed the staff in the laboratory on Leesburg Pike—which had been handling monkey blood and tissue samples—about the need to handle these specimens as if they were infected with the AIDS virus.

I must remember to inform labs that have received animal shipments from us to notify the C.D.C. if any unusual animal deaths occur. What about the exposure to those people who had been working on the air-handling system? What about the laundry service? Wasn't there a telephone repairman in recently? Perhaps last week—I can't remember just when that was. Holy Christ! Have I missed anything?

While he was updating the day's events on the computer, the telephone rang. It was Nancy Jaax on the line. She sounded tired. She told him that she had just finished the necropsies of the seven animals. She told him that her findings were consistent with either SHF or Ebola. She said it could be either one or both. Her results were ambiguous.

Reconnaissance

By the time Dan Dalgard woke up the next morning—it was now Thursday, exactly a week after Thanksgiving Day—he had made up his mind to invite the Army in to clean up one room, Room H, where the outbreak now seemed to be centered. He telephoned C. J. Peters and gave the Army permission to enter the monkey house. The news that they had the green light for a biohazard operation spread instantly through USAMRIID.

Colonel Jerry Jaax called a meeting of all the commissioned officers on his staff, along with two sergeants. They were Major Nathaniel ("Nate") Powell, Captain Mark Haines, Captain Steven Denny, Sergeant Curtis Klages, and Sergeant Thomas Amen, and he invited a civilian animal caretaker named Merhl Gibson to attend. These people were the core of his team. He put it casually to them: "Do you want to go to Reston?" Some of them had not heard of Reston. He explained what was going on, saying, "There are some monkeys that need to be euthanized. We'd like for you to play. Do you want in? Do you want to go?" They all said they wanted to play. He also figured that Nancy was going to play. That meant that he and Nancy would be inside the building at the same time. The children would be on their own tomorrow.

They were going to make an insertion into the monkey house, go into one room, kill the monkeys in that room, and take samples of tissue back to the Institute for analysis. They were going to do the job in space suits, under conditions of Level 4 biocontainment. The team would move out at 0500 hours tomorrow morning. They had less than twenty-four hours to get ready. Gene Johnson was gathering his biohazard equipment right now.

Gene drove down to Virginia and arrived at the monkey house in midmorning for a reconnaissance, to get a sense of the layout of the building and to figure out where to put the air lock and gray zone, and how to insert the team into the building. He went with Sergeant Klages, who was wearing fatigues. As they turned into the parking lot, they saw a television van parked in front of the monkey house, the newscaster and his crew drinking coffee and waiting for something to happen. It made Gene nervous. The news media had begun to circle around the story early on, but they couldn't seem to get a handle on it, and USAMRIID was trying to keep it that way.

Gene and the sergeant parked under a sweet-gum tree by the low brick building and went in through the front door. As they opened the door, the smell of monkey almost knocked them over. Whoa, Sergeant Klages thought, Whoa—we shouldn't even be in here without a space suit. The building stank of monkey. Something ugly was happening here. The whole god-damned place could be hot; every surface could be hot. The monkey workers had stopped cleaning the cages, because they did not want to go into the monkey rooms.

They found Bill Volt and told him they wanted to scout the building to determine the best way for the teams to enter tomorrow. Volt offered them a chair in his office while they talked. They didn't want to sit down, didn't want to touch any surfaces in his office with their bare hands. They noticed that Volt had a candy habit. He offered them a box full of Life Savers, Bit-O-Honeys, and Snickers bars—"Help yourselves," he said. Sergeant Klages stared at the candy with horror and mumbled, "No, thank you." He was afraid to touch it.

Gene wanted to go into the monkey area and see Room H, the hot spot. It was at the back of the building. He did not want to walk through the building to get to that room. He did not want to breathe too much of the building's air. Poking around, he discovered another route to the back of the building. The office space next door was empty and had been vacated some time ago; the electric power was cut off, and ceiling panels were falling down. He got a flashlight and circled around through these dark rooms. This is like a bombed-out area, he thought.

He found a door leading back into the monkey house. It led to a storeroom, and there was a closed corridor that headed deeper into the monkey house. Now he could see it all in his mind's eye. The closed corridor would be the air lock. The storeroom would be the staging area. The team could put on their space suits in this storeroom, out of sight of the television cameras. He drew a map on a sheet of paper.

When he understood the layout of the building, he circled to the front and told the monkey workers that he wanted the back areas of the building completely sealed off—airtight. He didn't want an agent from Room H to drift to the front of the building and get into the offices. He wanted to lower the amount of contaminated air flowing into those offices.

There was a door that led to the back monkey rooms. They taped it shut with military brown sticky tape: the first line of defense against a hot agent. From now on, as Gene explained to the monkey workers, no one was to break the sticky tape, no one was to go inside those back rooms except Army people until Room H had been cleaned out. What

Gene did not realize was that there was another way into the back rooms. You could get there without breaking the sticky tape on the door.

At eleven-thirty that morning, Lieutenant Colonel Nancy Jaax and Colonel C. J. Peters arrived at the corporate offices of Hazleton Washington on Leesburg Pike to meet with Dan Dalgard and to speak to a group of Hazleton lab workers who had been exposed to tissues and blood from sick monkeys. Since the C.D.C. now had charge of the human aspects of the Ebola outbreak, Joe McCormick also arrived at the Hazleton offices at the same time as Jaax and Peters.

The lab employees had been handling tissue and blood from the monkeys, running tests on the material. They were mainly women, and some of them were extremely frightened, nearly in a panic. That morning, there had been radio reports during rush hour, as the women were coming to work, that Ebola virus had killed hundreds of thousands of people in Africa. This was a wild exaggeration. But the radio newscasters had no idea what was going on, and now the women thought they were going to die. "We've been hearing about this on the radio," they said to Jaax and McCormick.

Nancy Jaax claims that Joe McCormick did his best to calm them down, but that as he talked to the women about his experiences with Ebola in Africa, they seemed to become more and more frightened.

A woman got up and said, "We don't care if he's been to Africa. We want to know if we're going to get sick!"

McCormick doesn't have any recollection of speaking to the women. He said to me, "I never talked to them. Nancy Jaax talked to them about Ebola."

Nancy thinks that they began to calm down when they saw a female Army colonel in a uniform. She asked the women, "Did any of you break a test tube? Do we have *anyone* here who stuck himself with a needle or cut himself?"

No one raised a hand.

"Then you'll be all right," she said to them.

A few minutes afterward, Dan Dalgard turned to C. J. Peters and, in a low voice, said something like, "Why don't you come over to the primate facility with me to look at the monkeys?"

Conspicuously, he did not invite McCormick of the C.D.C.

The Army was finally getting its foot in the door of the building.

They drove to the monkey house. By this time, Gene Johnson had closed off the back rooms and sealed the main entry door with sticky tape. Nancy and C. J., along with Dan Dalgard, circled around to the back of the building, put on rubber gloves and paper surgical masks, and went into Room H to look at the sick monkeys. Nancy and C. J. noticed with some concern that the monkey workers around the building were not wearing respirators, despite Dalgard's order. No one offered a respirator to Nancy or C. J. either. This made them both nervous, but they did not say anything. When in a monkey house, do as the monkey workers do. They did not want to give offense by asking for breathing equipment, not at this delicate moment, not when they had finally gotten their first chance to look at the building.

In Room H, Dalgard picked out the sick animals, pointing to them. "This one is sick, this one looks sick, this one over here looks sick," he said. The monkeys were quiet and subdued, but they rattled their cages now and then. Nancy stood well back from the cages and took shallow breaths, not wanting to let the smell of monkey get too deep into her lungs. A number of the animals had already died—there were many empty cages in the room—and many of the other animals were obviously sick. They sat at the backs of their cages, passive and blank faced. They were not eating their monkey biscuits. She saw that some had runny noses. She averted her eyes and behaved respectfully around the monkeys, because she did not want a monkey to get a notion in its head to spit at her. They have good aim when they spit, and they aim for your face. She worried more about her eyes than anything else. Ebola has a special liking for the eyes. Four or five virus particles on the eyelid would probably do it.

She noticed something else that made her fearful. These monkeys had their canine teeth. The company had not filed down the monkeys' fangs. The canines on these hummers were as big as the canines on any guard dog you'll ever see, and that was a rude awakening. A monkey can run amazingly fast, it can jump long distances, and it uses its tail as a gripper or a hook. It also has a mind. Nancy thought, An angry monkey is like a flying pit bull terrier with five prehensile limbs—these critters can do a job on you. A monkey directs its attacks toward the face and head. It will grab you by the head, using all four limbs, and then it will wrap its tail around your neck to get a good grip, and it will make slashing attacks all over your face with its teeth, aiming especially for the eyes. This is not a good situation if the monkey happens to be infected with Ebola virus. A six-foot-tall man and a ten-pound monkey are pretty evenly matched in a stand-up fight. The monkey will be all over the man. By the end of the fight, the man may need hundreds of stitches, and could be blinded. Jerry and his team would have to be exquisitely careful with these monkeys.

That evening, Jerry drove home alone. Nancy had put on a space suit and gone back into her lab to continue analyzing the monkey samples, and he had no idea when she would finish. He changed out of his uniform, and the telephone rang. It was Nancy's brother on the line, calling from Kansas, saying that Nancy's father was slipping, and that it looked as if the end was near. Nancy might be called home at any time for her father's funeral. Jerry said that he would pass the word along to Nancy, and explained that she was working late.

Then he and Jason drove for half an hour in the direction

of Washington and picked up Jaime at her gym. They decided to have supper at McDonald's. The Jaax family, minus
the mother, sat at a table, and while they ate, Jerry explained
to the children why Mom was working late. He said, ''Tomorrow morning, we're going to be going down to a civilian
place in space suits. There's an important thing going on
there. There are some monkeys that are sick. The situation
has kind of an emergency feel to it. We'll be gone real early,
and we may not get back until real late. You kids will be on
your own.'' They didn't react much to what he said.

Jerry went on, ''It's possible that humans could get sick
from the monkeys.''

''Well, there's not *really* any danger,'' Jaime said, chewing her chicken nuggets.

''Well, no, it's not really dangerous,'' he said. ''It's more
exciting than dangerous. And anyway, it's just what your
mom and I are doing right now.''

Jason said that he had seen something on television about
it. It was on the news.

''I think what your mom does is something pretty unusual,'' Jerry said to his son. And he thought, I'll never
convince him of that.

They returned home around nine-thirty, and Jerry had
trouble making the kids go to bed. Perhaps they were afraid
of what was happening but didn't know how to express it; he
wasn't sure. More likely, they sensed an opportunity to have
their own way when their mother wasn't around. They said
they wanted to wait up for her. He thought he would wait up
for her, too. He made them put on their pajamas, and he
brought them into bed with him, and they curled up on
Nancy's side of the water bed. There was a television in the
room, and he watched the eleven-o'clock news. A newscaster was standing in front of the monkey house, and he was
talking about people dying in Africa. By this time, the children had fallen asleep. He thought about his dead brother
John for a while, and then he picked up a book to try to read.

He was still awake when Nancy arrived home at one o'clock in the morning, looking fresh and clean, having taken a shower and shampooed her hair on her way out of Level 4.

As she looked around the house to see what needed to be done, she saw that Jerry had not tended to the animals. She put out food for the cats and dogs, and changed their water. She checked on Herky, the parrot, to see how he was doing. He started making noise the moment he perceived that the cats were being fed. He wanted some attention, too.

"Mom! Mom!" Herky hung upside down and laughed like a maniac, and cried, "Bad bird! Bad bird!" She took him out of his cage and stroked him on the head. He moved onto her shoulder, and she preened his feathers.

Upstairs in the bedroom, she found the children asleep next to Jerry. She picked up Jaime and carried her into her own bedroom and tucked her into bed. Jerry picked up Jason and carried him to his bed—he was getting too big for Nancy to haul around.

Nancy settled into bed with Jerry. She said to him, "I have a gut feeling they're not going to be able to contain the virus in that one room." She told him she was worried that it could be spreading into other rooms through the air. That virus was just so damned infective she didn't see how it would stay in one room. Something that Gene Johnson had once said to her came into her mind: "We don't really know what Ebola has done in the past, and we don't know what it might do in the future."

Then Jerry broke the news to her about her father. Nancy was beginning to feel extremely guilty about not going home to be with him as he lay dying. She felt the tug of her last obligation to him. She wondered if she should bag this monkey thing and fly to Kansas. But she felt that it was her duty to go through with the operation. She decided to take a chance that her father would live awhile longer.

PART THREE

SMASHDOWN

Insertion

DECEMBER 1, FRIDAY

The alarm went off at four-thirty. Jerry Jaax got up, shaved, brushed his teeth, threw on clothes, and was out of there. The teams were going to wear civilian clothes. No one wanted to attract attention. Soldiers in uniforms and camouflage, putting on\space suits . . . it could set off a panic.

It was five o'clock by the time he arrived at the Institute. There was no sign of dawn in the sky. A crowd of people had already gathered by a loading dock on the side of the building, under floodlights. There had been a hard freeze during the night, and their breath steamed in the air. Gene Johnson, the Ajax of this biological war, paced back and forth across the loading dock among a pile of camouflaged military trunks—his stockpile of gear from Kitum Cave. The trunks contained field space suits, battery packs, rubber gloves, surgical scrub suits, syringes, needles, drugs, dissection tools, flashlights, one or two human surgery packs, blunt scissors, sample bags, plastic bottles, pickling preservatives, biohazard bags marked with red flowers, and hand-pumped garden sprayers for spraying bleach on space suits and objects that needed to be decontaminated. Holding a cup of coffee in his fist, Gene grinned at the soldiers and rumbled, *"Don't touch my trunks."*

A white unmarked supply van showed up. Gene loaded his trunks into the van by himself and set off for Reston. He was the first wave.

By now, copies of *The Washington Post* were hitting driveways all over the region. It contained a front-page story about the monkey house:

Deadly Ebola Virus Found in Va.
Laboratory Monkey

One of the deadliest known human viruses has turned up for the first time in the United States, in a shipment of monkeys imported from the Philippines by a research laboratory in Reston.

A task force of top-level state and federal experts on contagious diseases spent much of yesterday devising a detailed program to trace the path of the rare Ebola virus and who might have been exposed to it. That includes interviews with the four or five laboratory workers who cared for the animals, which have since been destroyed as a precaution, and any other people who were near the monkeys.

Federal and state health officials played down the possibility that any people had contracted the virus, which has a 50 to 90 percent mortality rate and can be highly contagious to those coming into direct contact with its victims. There is no known vaccine.

"There's always a level of concern, but I don't think anybody's panicked," said Col. C. J. Peters, a physician and expert on the virus.

C. J. knew that if people learned what this virus could do, there would be traffic jams heading out of Reston, with mothers screaming at television cameras, "Where are my children?" When he talked to the *Washington Post* reporters, he was careful not to discuss the more dramatic aspects of the operation. ("I thought it would not be a good idea to talk about space suits," he explained to me much later.) He was careful not to use scary military terms such as *virus amplification, lethal chain of transmission, crash and bleed,* or *ma-*

jor pucker factor. A military biohazard operation was about to go down in a suburb of Washington, and he sure as hell didn't want the *Post* to find out about it.

Half of this biocontainment operation was going to be news containment. C. J. Peters's comments to *The Washington Post* were designed to create an impression that the situation was under control, safe, and not all that interesting. C. J. was understating the gravity of the situation. But he could be very smooth when he wanted, and he used his friendliest voice with the reporters, assuring them over the telephone that there really was no problem, just kind of a routine technical situation. Somehow the reporters concluded that the sick monkeys had been "destroyed as a precaution" when in fact the nightmare, and the reason for troops, was that the animals hadn't been destroyed.

As to whether the operation was safe, the only way to know was to try it. Peters felt that the larger danger could come from sitting back and watching the virus burn through the monkeys. There were five hundred monkeys inside that building. That was about three tons of monkey meat—a biological nuclear reactor having a core meltdown. As the core of monkeys burned, the agent would amplify itself tremendously.

C. J. arrived at the loading dock of the Institute at five o'clock in the morning. He would accompany the group down to the monkey house to see Jerry's team inserted, and then he would drive back to the Institute to deal with the news media and government agencies.

At six-thirty, he gave an order to move out, and the column of vehicles left Fort Detrick's main gate and headed south, toward the Potomac River. It consisted of a line of ordinary automobiles—the officers' family cars, with the officers inside wearing civilian clothes, looking like commuters. The line of cars followed behind two unmarked military vehicles. One was a supply van and the other was a snow-white ambulance. It was an unmarked Level 4 biocontainment ambulance. Inside it there were an Army medical-evacuation team and a biocontainment pod known as a bubble stretcher. This was a combat medical stretcher enclosed by a biocontainment bubble made of clear plastic. If someone was bitten by a monkey, he would go into the bubble, and from there he would be transferred to the Slammer. The supply van was a white unmarked refrigerator truck. This was to hold dead monkeys and tubes of blood.

There was not a uniform in the group, although a few members of the ambulance team wore camouflage fatigues. The caravan crossed the Potomac River at Point of Rocks and hit Leesburg Pike just as rush hour began. The traffic became bumper to bumper, and the officers began to get frustrated. It took them two hours to reach the monkey house, contending the whole way with ill-tempered commuters. Finally the column turned into the office park, which by that time was filling up with workers. The supply van and the ambulance were driven along the side of the monkey house, up onto a lawn, and were parked behind the building, to get them out of sight. The back of the building presented a brick face, some narrow windows, and a glass door. The door was the insertion point. They parked the supply van up close to the door.

At the edge of the lawn, behind the building, there was a line of underbrush and trees sloping down a hillside. Beyond that, there was a playground next to a day-care center. They could hear shouts of children in the air, and when they looked through the underbrush, they could see bundled-up four-year-olds swinging on swings and racing around a playhouse. The operation would be carried out near children.

Jerry Jaax studied a map of the building. He and Gene Johnson had decided to have all the team members put on their space suits inside the building rather than out on the lawn, so that if any television crews arrived there would be nothing to film. The men went through the insertion door and found themselves in an empty storage room. It was the staging room. They could hear faint cries of monkeys beyond a cinder-block wall. There was no sign of any human being in the monkey house.

Jerry Jaax was going to be the first man in, the point man. He had decided to take with him one of his officers, Captain Mark Haines, a former Green Beret. He was a short, intense man with a whipcord body who had been through the Green Berets' scuba-diving school. He had jumped out of airplanes at night into the open sea, wearing scuba gear. ("I'll tell you one thing," Haines once said to me. "I don't do scuba diving for fun as a civilian. The majority of my diving has been in the Middle East.") Captain Haines was not a man who would get claustrophobia and go into a panic in a space suit. Furthermore, Captain Haines was a veterinarian. He understood monkeys.

Jaax and Haines climbed into the supply van and pulled a plastic sheet across the van's back door for privacy, and stripped naked, shivering in the cold. They put on surgical scrub suits and then walked across the lawn, opened the glass door, and went into the storage room, the staging area, where an Army support team—the ambulance team, led by a captain named Elizabeth Hill—helped them into their space suits. Jerry knew nothing about field biological suits, and neither did Captain Haines.

The suits were orange Racal suits, designed for field use with airborne biological agents, and they were the same type of suit that had been used at Kitum Cave—in fact, some of them had come back from Africa in Gene Johnson's trunks. The suit has a clear, soft plastic bubble for a helmet. The suit is pressurized. Air pressure is supplied by an electric motor that sucks air from the outside and passes it through virus filters and then injects it into the suit. This keeps the suit under positive pressure, so that any airborne virus particles will have a hard time flowing into it. A Racal suit performs the same job as a heavy-duty Chemturion space suit. It protects the entire body from a hot agent, surrounding the body with superfiltered air. Army people generally don't refer to Racals as space suits. They call them Racals or field biological suits; but they are, in fact, biological space suits.

Jaax and Haines put on rubber gloves, and the support team taped the gloves to the sleeves of the suits while they held their arms out straight. On their feet, they wore sneakers, and over the sneakers they pulled bright yellow rubber boots. The support team taped the boots to the legs of the suits to make an airtight seal above the ankle.

Jerry was terribly keyed up. In the past he had lectured Nancy on the dangers of working with Ebola in a space suit, and now he was leading a team into an Ebola hell. At the moment, he didn't care what happened to himself, personally. He was expendable, and he knew it. Perhaps he could forget about John for a while in there. He switched on his electric blower, and his suit puffed up around him. It didn't feel too bad, but it made him sweat profusely. The door was straight ahead. He held the map of the monkey house in his hand and nodded to Captain Haines. Haines was ready. Jerry opened the door, and they stepped inside. The sound of the monkeys became louder. They were standing in a windowless, lightless, cinder-block corridor that had doors at either end: this was the makeshift air lock, the gray zone. The rule inside the air lock was that the two doors, the far door and the near door, could never be open at the same time. This was to prevent a backflow of contaminated air from coming into the staging room. The door closed behind them, and the corridor went dark. It went pitch-dark. Aw, son of a bitch. We forgot to bring flashlights. Too late now. They proceeded forward, feeling their way down the walls to the door at the far end.

Nancy Jaax woke up her children at seven-thirty. She had to shake Jason, as always, to get him out of bed. It didn't work, so she turned one of the dogs loose on him. He hit the bed flying and climbed all over Jason.

She put on sweatpants and a sweat shirt and went downstairs to the kitchen and flipped on the radio and tuned it to a rock-and-roll station and popped a Diet Coke. The music fired up the parrot. Herky began to scream along with John Cougar Mellencamp. Parrots really respond to electric guitar, she thought.

The children sat at the kitchen table, eating instant oatmeal. She told them that she would be working late, so they would be on their own at suppertime. She looked in the freezer and found a stew. It would do fine for the kids. They could defrost it in the microwave. She watched from the kitchen window as they walked down the driveway to the bottom of the hill to wait for the school bus. . . . "This work is not for a married female. You are either going to neglect your work or neglect your family" were the words of a superior officer long ago.

She cut a bagel for herself, and brought along an apple, and ate them in the car on the way to Reston. By the time she arrived at the monkey house, Jerry had already suited up and gone inside.

The staging room was crowded, warm, loud, confused. The experts on the use of space suits were giving advice to team members as they suited up. Nancy herself had never worn a Racal field suit, but the principles are the same as with a heavy-duty Chemturion. The main principle is that the interior of the space suit is a cocoon housing the normal world, which you bring with you into the hot area. If there were a break in the suit, the normal world would vanish, merging with the hot world, and you would be exposed. She spoke to the soldiers as they suited up. "Your suits are under pressure," she said. "If you get a rip in your suit, you have to tape it shut right away or you'll lose your pressure, and contaminated air could flow inside the suit." She held up a roll of brown sticky tape. "Before I go in, I wrap extra tape around my ankle, like this." She demonstrated how to do it: she wound the roll around her ankle several times, the way you tape up a sprained ankle. "You can tear off a length of tape from your ankle and use it to patch a hole in your suit," she said. "A hundred chancy things can happen to rip your suit."

She told them about Ebola in monkeys. "If these monkeys are infected with Ebola, then they are so full of virus that a bite from one of them would be a devastating exposure," she said. "Animals that are clinically ill with Ebola shed a *lot* of virus. Monkeys move real quick. A bite would be a death warrant. Be exquisitely careful. Know where your hands and body are at all times. If you get blood on your suit, stop what you are doing and clean it off right away. Don't let blood stay on your gloves. Rinse them off right away. With bloody gloves, you can't see a hole in the glove. Also, one other thing. You really don't want to drink a lot of coffee or liquids before going in. You will be in your space suit for a long time."

The batteries that pressurized the suit had a life span of six hours. People would have to leave the hot area and be deconned out before their batteries failed, or they would be in trouble.

Jerry Jaax and Captain Mark Haines felt their way down the dark corridor, toward the door that led into the hot zone. They opened it and found themselves standing at the junction of two corridors, bathed in a cacophony of monkey cries. The air-handling equipment still wasn't working, and the temperature in the place seemed as if it was above ninety degrees. Jerry's head bubble fogged up. He pushed the bubble against his face to rub the moisture off the faceplate, and now he could see. The walls were gray cinder block, and the floor was painted concrete.

Just then, he noticed a blur of motion on his left, and he

turned and saw two Hazleton workers walking toward him. They weren't supposed to be in here! The area was supposed to be sealed off, but they had come in by another route that led through a storeroom. They wore respirators, but nothing covered their eyes. When they saw the two men in space suits, they froze, speechless. Jerry could not see their mouths, but he could see their eyes, wide with astonishment. It was as if they had suddenly discovered that they were standing on the moon.

Jerry didn't know what to say. Finally he said, "WHICH WAY TO ROOM H?"—shouting to be heard over his blowers.

The workers led him down the corridor to the infected room. It was at the far end of the hall. Then they retreated to the front of the building and found Dan Dalgard, who had been sitting in an office, waiting for the Army to come in. He showed up at Room H moments later, wearing a respirator, to find out what was going on. Jerry looked at him as if he was insane. It was as if you went to a meeting with someone and the person showed up naked.

Dalgard was not happy with the space suits. Apparently he had not realized how the Army would be outfitted. He gave them a tour of Room H, feeling exceptionally nervous. "Looks like we have some sick monkeys in here," he said. Some of the monkeys went berserk when they saw the space suits. They spun in circles in their cages or cowered in the corners. Others stared at the humans with fixed expressions on their faces.

"You see the clinical signs," Dalgard said, pointing to a monkey. "I feel pretty confident I can tell when a monkey is getting sick. They get a little bit depressed, they go off their feed, and in a day or two they are dead."

Jerry wanted to look at all the monkeys in the monkey house. He and Captain Haines went back out into the corridor and went from room to room through the entire building. They found other monkeys that seemed depressed, with the same glazed expression on their faces. Jaax and Haines, both of whom knew a lot about monkeys, didn't like the feel of this whole building. Something lived in here other than monkeys and people.

Nancy Jaax got ready to go inside. She changed into a scrub suit in the van, ran across the lawn, and entered the staging area. The support team helped her suit up. She gathered several boxes of syringes and went in with Captain Steven Denny. They walked down the air-lock corridor and came to the far door. She opened the door and found herself in the long corridor. It was empty. Everyone was down the hall in Room H. Jerry thought his wife looked like the Pillsbury dough boy. Her suit was too large for her, and it billowed around her when she walked.

Nancy noticed mucus and slime on the noses of some of the monkeys. That scared her, because it seemed so much like the flu or a cold, when it wasn't. Dan Dalgard, wearing a respirator and a jumpsuit, selected four sick monkeys for sacrifice, the ones he thought looked the sickest. He reached into the cages and gave the monkeys their shots. When they crumpled and fell asleep, he gave them a second round of shots, and that stopped their hearts.

The room was jammed with people in space suits. They kept coming in in pairs, and they milled around with nothing to do. One of them was Sergeant Curtis Klages. He turned to someone and said, "WELL, THIS IS A BIG CHARLIE FOXTROT." That's code for C.F., which means "cluster fuck." A Charlie Foxtrot is an Army operation that winds up in confusion, with people bumping into one another and demanding to know what's going on.

Nancy happened to glance at the sergeant, checking his suit instinctively, and she saw that he had a tear across his hip. She touched the sergeant's arm and pointed. She reached down to her ankle, where she kept her extra tape, and taped the hole for him.

She removed the four dead monkeys from their cages, holding them by the backs of the arms, and loaded them into plastic biohazard bags. She carried the bags to the entry door, where someone had left a garden sprayer full of Clorox bleach along with more bags. She double-bagged the monkeys, spraying each bag with bleach, and then she loaded the bags into cardboard biohazard containers—hatboxes—and sprayed them to decon them. Finally she loaded each hatbox into a third plastic bag and sprayed it. She pounded on the door. "IT'S NANCY JAAX. I'M COMING OUT." The door was opened by a sergeant standing on the other side, a member of the decon team. He was wearing a Racal suit, and he had a pump sprayer filled with bleach. She went into the air lock, pushing the hatboxes ahead of her.

In the darkness and in the whine of their blowers, he shouted to her, "STAND WITH YOUR ARMS OUT, AND TURN AROUND SLOWLY." He sprayed her for five minutes, until the air lock stank of bleach. It felt wonderfully cool, but the smell leaked through her filters and made her throat sting. He also sprayed the bags. Then he opened the door to the staging area, and she blinked at the light and came out, pushing the bags ahead of her.

The support team peeled off her suit. She was drenched with sweat. Her scrubs were soaked. Now it was freezing cold. She ran across the lawn and changed into her civilian clothes in the back of the van.

Meanwhile, people loaded the bags into boxes, and loaded the boxes into the refrigerator truck, and Nancy and a driver headed off for Fort Detrick. She wanted to get those monkeys into Level 4 and opened up as fast as possible.

Jerry Jaax counted sixty-five animals in the room, after the four that Nancy had removed. Gene Johnson had brought a special injector back from Africa. Jerry used this device to give shots to the monkeys. It was a pole that had a socket on one end. You fitted a syringe into the socket, and you slid the pole into the cage and gave the monkey a shot. You also needed a tool to pin the monkey down, because monkeys don't like needles coming at them. They used a mop handle with a soft U-shaped pad on the end. Captain Haines held the mop handle against the monkey to immobilize it, and Jerry ran the pole into the cage and hit the monkey's thigh with a double dose of ketamine, a general anesthetic. They went through the room from cage to cage, hitting all the monkeys with the drug. Pretty soon the monkeys began to collapse in their cages. Once a monkey was down, Jerry gave it a shot of a sedative called Rompun, which put it into a deep sleep.

When all the monkeys were down and asleep, they set up a couple of stainless-steel tables, and then, one monkey at a time, they took blood samples from the unconscious monkeys and gave them a third injection, this time of a lethal drug called T-61, which is a euthanasia agent. After a monkey was clinically dead, it was opened up by Captain Steve Denny. He took samples of liver and spleen, using scissors, and dropped the samples into plastic bottles. They bagged the dead monkeys, loaded them into hatboxes, and piled the hatboxes along the corridor. Dan Dalgard, meanwhile, left the room and remained in an office at the front of the building for the rest of the day.

By late afternoon, all the monkeys in Room H had been put to death. Behind the building, through the trees and down the hill, children ran in circles around their playhouse. Their shouts carried far in the December air. Their mothers and fathers arrived in cars and picked them up. The team exited from the hot zone in pairs, and stood around on the grass in their civilian clothes, looking pale, weak, and thoughtful. In the distance, floodlights began to light up the monuments and buildings of Washington. It was the Friday evening at the end of the week following Thanksgiving, the start of a quiet weekend that precedes the Christmas season. The wind strengthened and blew paper cups and empty cigarette packs in eddies around the parking lots. In a hospital not far from there, Jarvis Purdy, the monkey worker who had had a heart attack, rested comfortably, his condition stable.

Back at the Institute, Nancy Jaax again stayed up until one o'clock in the morning, dissecting monkeys with her hot-zone buddy, Ron Trotter. When they had suited up and gone in, there had been five monkey carcasses waiting for them in the air lock.

This time, the signs of Ebola were obvious. Nancy saw what she described as "horrendous gut lesions" in some of the animals, caused by sloughing of the intestinal lining. That sloughing of the gut was a classic sign. The intestine was blitzed, completely full of uncoagulated, runny blood, and at the same time the monkey had had massive blood clotting in the intestinal muscles. The clotting had shut off blood circulation to the gut, and the cells in the gut subsequently died—that is, the intestines had died—and then the gut had filled up with blood. Dead intestine—this was the kind of thing you saw in a decayed carcass. In her words, "It looked like the animals had been dead for three or four days." Yet they had been dead only for hours. Some of the monkeys were so badly liquefied that she and Trotter didn't even bother to do a necropsy, they just yanked samples of liver and spleen from the dead animal. Some of the monkeys that were dying in Room H had become essentially a heap of mush and bones in a skin bag, mixed with huge amounts of amplified virus.

DECEMBER 4, 0730 HOURS, MONDAY

Monday arrived cold and raw, with a rising wind that brought a smell of snow from a sky the color of plain carbon steel. In the shopping malls around Washington, Christmas lights had been hung. The parking lots were empty, but later in the day they would fill up with cars, and the malls would fill up with parents and children, and the children would line up to see Santa Clauses. Dan Dalgard drove to the primate building, one more commuter in a sea of morning traffic.

He turned into the parking lot. As he got closer to the building, he saw that a man was standing by the front door near the sweet-gum tree, wearing a white Tyvek jumpsuit. It was one of the monkey caretakers. Dalgard was furious. He had instructed them not to come out of the building wearing a mask or a protective suit. He jumped out of his car, slammed the door, and hurried across the parking lot. As he got closer, he recognized the man as someone who will be called Milton Frantig. Frantig was standing bent over with his hands on his knees. He didn't seem to notice Dalgard—he was staring at the grass. Suddenly Frantig's body convulsed, and liquid spewed out of his mouth. He vomited again and again, and the sound of his retching carried across the parking lot.

A Man Down

As Dan Dalgard watched the man spill his stomach out onto the lawn, he felt, in his words, "scared shitless." Now, perhaps for the first time, the absolute horror of the crisis at the primate building washed over him. Milton Frantig was doubled over, gasping and choking. When his vomiting subsided, Dalgard helped him to his feet, took him indoors, and made him lie down on a couch. Two employees were now sick—Jarvis Purdy was still in the hospital, recovering from a heart attack. Milton Frantig was fifty years old. He had a chronic, hacking cough, although he didn't smoke. He had been working with monkeys and with Dalgard at Hazleton for more than twenty-five years. Dalgard knew the man well and liked him. Dalgard felt shaken, sick with fear and guilt. *Maybe I should have evacuated the building last week. Did I put the interests of the monkeys ahead of the interests of the human beings?*

Milton Frantig was pale and shaky, and felt faint. He developed the dry heaves. Dalgard found a plastic bucket for him. Between heaves, interrupted by coughing spells, Frantig apologized for leaving the building while wearing a jump suit. He said he had just been putting on his respirator to go inside a monkey room when he began to feel sick to his stomach. Perhaps the bad smell in the building had nauseated him, because the monkey rooms weren't being cleaned as regularly as usual. He could feel he was about to vomit, and he couldn't find a bucket or anything to throw up into, and it was coming on so fast that he couldn't get to the rest room, so he had run outdoors.

Dalgard wanted to take Frantig's temperature, but nobody could find a thermometer that hadn't been used rectally on monkeys. He sent Bill Volt to a drugstore to buy one. When he returned, they discovered that Frantig had a fever of a hundred and one. Bill Volt hovered in the room, almost shaking with fear. Volt was not doing well—"almost spastic in his terror," Dalgard would later recall, but it wasn't any different from the way Dalgard felt.

Milton Frantig remained the calmest person in the room. Unlike Dalgard and Volt, he did not seem afraid. He was a devout Christian, comfortable with telling people that he had been saved. If the Lord saw fit to take him home with a monkey disease, he was ready. He prayed a little, remembering his favorite passages in the Bible, and his dry heaves subsided. Soon he was resting quietly on the couch and said he felt a little better.

"I want you to stay where you are," Dalgard said to him. "Don't leave the building." He got into his car and drove as fast as he could to the Hazleton Washington offices on Leesburg Pike. The drive didn't take long, and by the time he got there, he had made up his mind: the monkey facility had to be evacuated. Immediately.

There had been four workers employed in the building, and two of them were now going to be in the hospital. One man had heart problems, and now the other had a fever with vomiting. From what Dalgard knew about Ebola virus, either of these illnesses could be signs of infection. They had shopped at malls and visited friends and eaten in restaurants. Dalgard thought they were probably having sexual intercourse with their wives. He didn't even want to think about the consequences.

When he arrived at Hazleton Washington, he went directly

to the office of the general manager. He intended to brief him about the situation and get his approval to evacuate the monkey house. "We've got two guys who are sick," Dalgard said to him. He began to describe what had happened, and he started crying. He couldn't control it. He broke down and wept. Trying to pull himself together, he said, "I recommend that—the entire operation—be shut down—as soon as possible. My recommendation is—we close it down and turn it over to the Army. We've had this god-damned disease since October, we haven't gotten injured, and all of a sudden we've got two guys sick, one in the hospital, one who's going there. I kept on thinking that if there was a real human risk, we would have seen something by now. We've played with fire for too long."

The general manager sympathized with Dalgard and agreed with him that the monkey facility ought to be evacuated and shut down. Then, holding back his tears, Dalgard hurried to his own office, where he found a group of officials from the C.D.C. waiting for him. He felt as if the pressure would never let up. The C.D.C. people had arrived at Hazleton to begin surveillance of all employees who had been exposed to the virus. Dalgard told them what had just happened at the monkey house, that a man had gone down with vomiting. He said, "I have recommended that the facility be evacuated. I feel that the building and the monkeys should be turned over to the people from USAMRIID, who have the equipment and personnel to handle it safely."

The C.D.C. people listened and did not disagree.

Then there was the question of what to do with Milton Frantig, who was still lying on the couch at the monkey house under orders from Dalgard not to move. Since the C.D.C. was in charge of the human aspects of the outbreak, the C.D.C. was in charge of Frantig—and the C.D.C. wanted him taken to Fairfax Hospital, inside the Washington Beltway.

It was now nine-twenty in the morning. Dalgard sat in his

office and sweated it out, managing the crisis by telephone. He called C. J. Peters at Fort Detrick and told him that he had a monkey caretaker who was sick. In his dry, calm voice, now without any hint that he had recently been weeping, he said to Peters, "You have permission to consider the facility and all the animals to be the responsibility of USAMRIID."

Colonel C. J. Peters was shocked to hear that a man had gone down, but he was a little distrustful of the phrase "the responsibility of USAMRIID." It implied that if anything went wrong and people died, the Army could be held responsible and could be sued. He wanted to take control of the building and sterilize it, but he didn't want lawsuits. So he said to Dalgard that the safety of his people and the safety of the general public were the most important things to him but that he would have to clear this with his command. He said he would get back to Dalgard as soon as possible.

Then they talked about the sick man, and C. J. learned that he was being taken to Fairfax Hospital. That disturbed him greatly. He felt that it should be assumed that the guy was breaking with Ebola—and do you really want to bring a guy like that into a community hospital? Look at what Ebola had done in hospitals in Africa. Ebola could shut down a hospital; it could amplify itself in a hospital. C. J. thought the man belonged in the Slammer at the Institute.

As soon as he got off the line with Dalgard, C. J. Peters telephoned Joe McCormick, who was in charge of the C.D.C. effort, to try to persuade him to let the Army put the man in the Slammer. He said to McCormick something like, "I know you have this idea that a surgical mask and gown are all you need to handle an Ebola patient, but I think you need to use a higher level of containment," and he offered to pick up the sick man in an Army ambulance—put him in an Army biocontainment pod—and carry the pod to the Army's facilities at the Institute. Put him in the Slammer.

C. J. Peters recalls that McCormick said to him something

like, "I want the guy at Fairfax Hospital." C. J. replied, "All right. I believe *this,* Joe, and you believe *that,* and we don't agree. Regardless—what is going to happen to the medical personnel at Fairfax Hospital or to you, Joe, if Ebola virus gets into that hospital?"

McCormick would not budge on his decision: he had been face-to-face with Ebola in Africa and he hadn't gotten sick. He had worked for days inside a mud hut that was smeared with Ebola blood, on his knees among people who were crashing and bleeding out. You didn't need a space suit to handle an Ebola patient. They could be handled by skilled nurses in a good hospital. The guy was going to Fairfax Hospital. C. J. Peters, in spite of his strong dislike for McCormick, found himself admiring him for making strong decisions in a very difficult situation.

At this moment, a television-news van arrived at the monkey house from Channel 4 in Washington. The workers peered through curtains at the van, and when the reporter came to the door and pushed the buzzer, no one answered. Dalgard had made it clear to them that no one was to talk to the media. Just then, an ambulance from Fairfax Hospital arrived to take Frantig away. Channel 4's timing could not have been better. The news team turned on their lights and started filming the action. The door of the monkey house swung open and Milton Frantig stumbled out, still wearing his Tyvek suit, looking embarrassed. He walked over to the ambulance, the medical team opened the back doors of the vehicle, and Frantig climbed in by himself and lay down on the gurney. They slammed the doors and took off with Channel 4 following them. A few minutes later, the ambulance and Channel 4 pulled into Fairfax Hospital. Frantig was put in an isolation room, with access restricted to doctors and nurses wearing rubber gloves, gowns, and surgical masks. He said he felt better. He prayed to the Lord and watched a little television.

Back at the monkey house, the situation had become unbearable for the remaining workers. They had seen people in
space suits, they had seen their colleague puking in the
grass, they had seen Channel 4 chasing the ambulance. They
left the building in a real hurry, locking it after themselves.

There were four hundred and fifty monkeys alive in the
building, and their hoots and cries sounded in the empty
hallways. It was eleven o'clock in the morning. A snow
flurry came and went. The weather was turning colder. In the
monkey house, the air-handling equipment had failed for
good. The air temperature in the building had soared beyond
ninety degrees, and the place had turned steamy, odorous,
alive with monkey calls. The animals were hungry now, because they had not been fed their morning biscuits. Here and
there, in rooms all over the building, some of the animals
stared from glazed eyes in masklike faces, and some of them
had blood running from their orifices. It landed on metal
trays under their cages . . . *ping, ping, ping.*

91-Tangos

Dan Dalgard felt he was losing control of everything. He set up a conference call with all the senior managers in his company and informed them of the situation—two employees were down, and the second man could be breaking with Ebola—and he told the managers that he had offered to turn the monkey house over to the Army. They approved his action, but they said they wanted the oral agreement with the Army to be put in writing. Furthermore, they wanted the Army to agree to take legal responsibility for the building.

Dalgard then called C. J. Peters and asked that the Army assume responsibility for any liability that would arise after the Army took over. C. J. flatly rejected that proposal. He saw a need for clarity, speed, and no lawyers. He felt that the outbreak had ballooned to the point where a decision had to be made. Dalgard agreed to fax him a simple letter turning the monkey house over to the Army. They worked up some language, and C. J. carried the letter by hand to the office of General Philip Russell. He and the general pored over the letter, but they did not choose to show it to any Army lawyers. Russell said, "We have to convince the lawyers of the path of righteousness." They signed the letter, faxed it back to Dalgard, and the monkey house fell into the hands of the Army.

Jerry Jaax would have to lead a much larger biohazard team back into the monkey house. The number of animals that needed to be dealt with was staggering. His troops were untested, and he himself had never been in combat. He didn't know, couldn't know, how he or his people would perform in a chaotic situation involving intense fear of an unpleasant death.

Jerry was the commanding officer of the 91-Tangos at the Institute. The Army's animal-care technicians are classified 91-T, which in Army jargon becomes 91-Tango. The younger 91-Tangos are eighteen years old and are privates. While the ambulance was taking Milton Frantig to the hospital, Jerry called a meeting of his 91-Tangos and civilian staff in a conference room in the Institute. Although most of the soldiers were young and had very little or no experience in space suits, the civilians were older men, and some were Level 4 specialists who had worn Chemturions on a daily basis. The room was jammed, and people sat on the floor.

"The virus is Ebola or an Ebola-like agent," he said to them. "We are going to be handling large amounts of blood. And we will be handling sharp instruments. We are going to use the disposable biocontainment suits."

The room was silent while he spoke. He didn't mention that a man was down, because he didn't know about it—C. J. Peters hadn't told him about that. For the time being, Peters was staying quiet about that development.

Jerry said to his people, "We are looking for volunteers. Is there anyone in this room who *does not* want to go? We can't make you go."

When no one backed out, Jerry looked around the room and picked his people: "Yep, he's going. She's going, and, yep, you're going." In the crowd, there was a sergeant named Swiderski, and Jerry decided that she could not go because she was pregnant. Ebola has particularly nasty effects on pregnant women.

No combat unit in the Army could handle this work. There would be no hazard pay, as there is in a war zone. The Army has a theory regarding biological space suits. The theory is that work inside a space suit is not hazardous, because you are wearing a space suit. Hell, if you handled hot agents without a space suit, that would be hazardous work. The privates would get their usual pay: seven dollars an hour. Jerry told them that they were not to discuss the operation with anyone, not even members of their families. "If you have any tendency to claustrophobia, consider it now," he said. He told them to wear civilian clothes and to show up at the Institute's loading dock at 0500 hours the next morning.

DECEMBER 4–5, MONDAY–TUESDAY

The soldiers didn't sleep much that night, and neither did Gene Johnson. He was terrified for the "kids," as he called them. He had had his fair share of scares with hot agents. Once in Zaire, he had stuck himself with a bloody needle while taking blood from a mouse. There was reason to believe the mouse was hot with Lassa (a Level 4 agent), and so they had airlifted him to the Institute and put him in the Slammer for thirty days. "That was not a fun trip," as he put it. "They treated me as if I would die. They wouldn't give me scissors to cut my beard because they thought I would be suicidal. And they locked me in at night." At Kitum Cave, while wearing a space suit and dissecting animals, he had been nicked three times with bloody tools. Three times his space suit had been punctured and his skin broken and the cut smeared with animal blood. He regarded himself as lucky not to have picked up Marburg or something else at Kitum Cave. Having had some close calls, he was deeply afraid of what had invaded the monkey house.

Johnson lived in a rambling house on the side of Catoctin
Mountain. He sat in his study most of the night, thinking
about procedures. Every movement of the body in a hot area
has to be controlled and planned. He said to himself,
Where's this virus going to get you? It's going to get you
through the hands. The hands are the weak point. Above all,
the hands must be under control.

He sat in an easy chair and held up one hand and studied
it. Four fingers and an apposed thumb. Exactly like a monkey's hand. Except that it was wired to a human brain. And
it could be enclosed and shielded by technology. The thing
that separated the human hand from Nature was the space
suit.

He stood up and went through motions in the air with his
hands. Now he was giving a monkey an injection. Now he
was carrying the monkey to a table. He was putting the
monkey on the table. He was in a hot zone. He was opening
the monkey up, and now he was putting his hands into a
bloody lake of amplified hot agent. His hands were covered
with three layers of rubber and then smeared with blood and
hot agent.

He paused and jotted notes on paper. Then he turned back
to his imaginary hot zone. He inserted a pair of scissors into
the monkey and clipped out part of the spleen. He handed it
to someone. Where would that person be standing? Behind
him? Now he imagined himself holding a needle in his hand.
Okay, I have a needle in my hand. It's a lethal object. I'm
holding it in my right hand if I'm right-handed. Therefore,
my buddy should stand to my left, away from the needle.
Now my buddy's hands. What will my buddy's hands be
doing? What will everyone's hands be doing? By early in the
morning, he had written many pages of notes. It was a script
for a biohazard operation.

Jerry Jaax left home at four o'clock in the morning, while
Nancy was still asleep. He met Gene Johnson at the loading

dock, where they went over Gene's script. Jerry studied it, and meanwhile the team members began to show up, soldiers in Jerry's unit. Many of them arrived on foot, having walked over from their barracks. They stood around, waiting for their orders. It was pitch-dark, and only the floodlights illuminated the scene. Jerry had decided to use a buddy system inside the building, and he began deciding who would be paired with whom. On a piece of paper, he drew up a roster of buddies, and he wrote down the order of entry, the sequence in which they would be inserted into the building. He stood before them and read the roster, and they got into their vehicles—a white refrigerator truck, a couple of unmarked passenger vans, an unmarked pickup truck, the white ambulance containing the bubble stretcher, and a number of civilian cars—and headed for Reston. They became trapped in rush-hour traffic again, surrounded by half-asleep yuppies in suits who were sucking coffee from foam cups and listening to traffic reports and easy rock and roll.

When all the vehicles had arrived at the back of the monkey house, the teams assembled on the lawn, and Gene Johnson asked for their attention. His eyes were sunken and dark, suggesting he had not slept in days. ''We are not playing games here,'' he said. ''This is the real thing. A biological Level 4 outbreak is not a training session. There's been a development I want you all to know about. There is a possibility that transmission of this virus to humans has taken place. There are two people who are ill and are hospitalized. Both of them are animal caretakers who worked in this building. There is one guy we are especially worried about. Yesterday morning, he became sick. He vomited and spiked a fever. He is now in the hospital. We don't know if he is breaking with Ebola. The thing I want you to understand is that he was not bitten by an animal and he did not cut himself or stick himself with a needle. So if he has Ebola, there is a possibility he got it through airborne transmission.''

Jerry Jaax listened to the speech with a rising sense of horror. He hadn't known about this man getting sick! Nobody had told him about this! Now he had a feeling that there were going to be casualties.

It was an icy, gray day. The trees behind the monkey house had lost their leaves, and dead leaves rustled across the lawn. At the day-care center down the hill, parents had been dropping off their children, and the children were playing on swings. Gene Johnson continued his speech. "Everyone is to proceed on the assumption that Ebola virus is potentially airborne," he said. "You know the risks, and you are experienced"—and his eyes rested on a private first class named Nicole Berke. She was quite beautiful, long blond hair, eighteen years old—and he thought, Who is she? I've never seen her before. Must be one of Jerry's people. They're just kids, they don't know what they're up against. "You must follow the procedures exactly," he went on. "If you have any questions, you must ask."

Jerry got up and said to them, "No question is too stupid. If you have a question, ask."

Private Nicole Berke was wondering if she would get a chance to go into the building. "How long are we going to be doing this, sir?" she asked him.

"Until the monkeys are dead," he replied. "There are four hundred and fifty monkeys in there."

Oh, God, she thought, *four hundred fifty monkeys*—this is going to take forever.

The questions were few. People were tense, silent, turned inward. Jerry Jaax entered the staging room, and the support team helped him put on his Racal suit. They fitted the bubble over his head, and his blowers started to roar. He told the teams he'd see them inside, and he and his buddy, Sergeant Thomas Amen, entered the air lock. The door closed behind them, and they stood in darkness. They felt their way down the dark air-lock corridor, opened the far door, and crossed over to the hot side.

The area was trashed. It had not been cleaned in many days. The workers had left in a big hurry. There were monkey biscuits scattered all over the floor, and papers were scattered everywhere, and there were overturned chairs in the offices. It looked as if the humans had fled from here. Jerry and the sergeant began exploring the building. They moved slowly and carefully in their suits, as if they were wreck divers operating in deep water. Jerry found himself in a small corridor that opened into more monkey rooms. He saw a room full of monkeys, and every one of the animals was looking out at him. Seventy pairs of monkey eyes fixed on a pair of human eyes in a space suit—and the animals went nuts. They were hungry and were hoping to be fed. They had trashed their room. Even locked inside cages, monkeys could really do a job on a room. They had been throwing their biscuits all over the place, and they had been finger-painting the walls with dung. The walls were scribbled all the way up to the ceiling with monkey writing. It was a cryptic message to the human race that came out of the primate soul.

Jerry and the sergeant found some bags of monkey biscuits, and went into each room in the building, and fed the monkeys. The animals were going to die soon, but Jerry didn't want them to suffer more than they had to. While he fed them, he inspected them for signs of Ebola. In many of the rooms, he found animals that seemed dull-eyed and listless. Some of them had runny noses, or there was a kind of blood-spattered green crust caked around the nostrils. He saw puddles of blood in some of the pans under their cages. These sights disturbed him deeply because they told him that the agent had gone all through the building. He could see some of the animals coughing and sneezing, as if they had the flu. He wondered if he was seeing a mutant form of Ebola—a kind of airborne Ebola flu. He shrank from the idea and tried to turn his mind away from it, for it was too awful to contemplate. You could no more imagine a season of Ebola flu than you could imagine a nuclear war. A layer of sweat built up inside his plastic head bubble, making it difficult for him to see the monkeys clearly. But he could hear them, shrieking and calling distantly beyond the sound of his blowers. So far, he had not felt any claustrophobia or panic. He wasn't going to lose it in here.

Several members of the team spent the next half hour in the staging room. They were shucking syringes, removing them from a sterile envelope, and fitting each syringe with a needle. Now the syringes were ready to be filled with drugs.

A few feet from the soldiers, Captain Mark Haines began to suit up. While the support team got him dressed, he gave a speech. He wanted the soldiers to keep certain things in mind as they followed him in. He said, "You are going to euthanize a whole building full of animals. This is not a fun operation. Don't get attached to the animals. They were going to die of Ebola anyway. They're all going to have to go, every last one of them. Don't think of it as killing something. Think of it as stopping the virus here without letting it get anywhere else. Don't play with the monkeys. I don't want to hear laughing and joking around the animals. I can be hard. Remember the veterinarian's creed: you have a responsibility to animals, and you have a responsibility to science. These animals gave their lives to science. They were caught up in this thing, and it's not their fault. They had nothing to do with it. Keep an eye on your buddy. Never hand a used needle to another person. If a needle comes out of its cap, it goes straight into an animal. And then put the used syringe immediately into a sharps container. If you get tired, tell your supervisor, and we'll decon you out." He turned away from them and went in with his buddy.

"Who's next?" Gene Johnson said, reading the roster. "Godwin! You're next."

A private first class named Charlotte Godwin hurried outdoors to the van and climbed inside, and took off all her clothes, and put on a surgical scrub suit, socks, sneakers, and a hair cap. It was brutally cold in the van. She felt embarrassed and vulnerable.

In the staging room, they began to suit her up. Someone said to her, "You're kind of small. We've got a special suit for you." It wasn't special. It was a large suit, sized for a big man, and she was five feet tall. It hung around her like a bag. The support team was taping her now, running brown sticky tape around her ankles and wrists, and her blowers came on.

An Army photographer took some photographs for the action file, and as the flash went off she thought, God, I would be wearing a hair cap. It's a clown cap. It's a Bozo hat. You won't see my hair in the picture, and my space suit is too large for me. Makes me look fat. Just my luck to be the one looking like a dork in the action photographs.

She staggered into the gray zone, carrying boxes of supplies, and felt a sweeping adrenal rush, and thought, I'm too young to be going through this. She was eighteen. Then she noticed the smell. A really bad smell was creeping through her filters. Her buddy pounded on the far door, and they entered. Ripples in the faceplate of her head bubble distorted the view, as if she was in a house of mirrors. The smell of monkey was overpowering inside her space suit. It was also too quiet, and monkey houses are not quiet places. The quiet bothered her even more than the smell or the heat.

A door swung open, and Colonel Jaax appeared. He said, "START LOADING SYRINGES. DOUBLE DOSES OF KETAMINE."

"YES, SIR," she replied.

"THE SERGEANT AND I WILL BE KNOCKING DOWN MONKEYS IN HERE," he said.

Charlotte started filling syringes with ketamine, the anesthetic. Jerry Jaax carried a loaded syringe into the monkey room and fitted it to the socket of a pole syringe. The sergeant fished his mop handle into a cage and pinned a monkey. Then Jerry opened the door of the cage. Watching the monkey carefully to make sure it didn't try to rush at him, he slid the pole syringe through the open door and gave the monkey an injection of anesthetic, and then pulled out the syringe, and slammed the door shut. It was the most dangerous job because of the open door. The animal could attack or try to escape. Jerry and the sergeant went from cage to cage, and the monkeys began to go to sleep under the anesthesia.

The rooms contained double banks of monkey cages. The lower bank was near the floor and was dark. Jerry had to get down on his knees to peer inside. He could hardly see anything through his head bubble. His knees were killing him. He would open a cage door, and the sergeant would slide the mop handle into the cage. The monkey would scrabble around, trying to escape, and the sergeant would say, "OKAY, I GOT HIM. HE'S PINNED." Jerry would slide the pole syringe toward the monkey, aiming the needle for the thigh. There would be screeches and a wild commotion, the monkey shrieking *"Kra! Kra!"* and the needle would sink in. This was turning out to be one of the hardest things he'd ever done in his career as a veterinarian.

More team members came into the building. Jerry assembled them in the hallway and said to them: "STOP EVERY FIVE OR TEN MINUTES AND CHECK YOUR NEIGHBOR'S SUIT FOR RIPS. BE VERY CAREFUL. MAKE SURE YOU TAKE REST BREAKS. I WANT YOU TO REST FOR TEN MINUTES EVERY HOUR. WHEN YOU GET TIRED, YOU GET CARELESS." Every time he looked into a monkey room, he saw a room full of eyes looking back at him. Some of the monkeys rattled their cages, and the wave of noise swept up and down the room.

Jerry decided to set up a bleed area in a small room near the front of the building, right next to the offices. In the bleed area there was a shower with a drain hole in the floor. They would need to use the drain hole for washing down blood and for rinsing objects with bleach. Every time blood went down the drain, they would pour bleach after it—they didn't want Ebola getting into the Reston sewage system. They found a metal examination table on wheels and rolled it into the bleed area. Jerry divided his people into subteams: a bleed team (to work at the bleed table), a euthanasia team (to put monkeys to death), and a necropsy team (to open up the monkeys and take samples and bag the carcasses in biohazard bags).

They got an assembly line going. Every five minutes or so, Jerry Jaax would carry an unconscious monkey out of a room and down the corridors to the bleed area, holding the animal with its arms pinned behind its back. He would lay it down on the bleed table, and then Captain Haines, the Green Beret, would insert a needle into the animal's thigh and draw off a lot of blood into various tubes. Then he'd hand the unconscious animal to Major Nate Powell, and he would give it an injection of T-61, the euthanasia agent. He'd put the needle right into the heart. When the animal was clearly dead, he would hand it to Captain Steve Denny, who did the necropsy. Captain Denny opened the animal with scissors and snipped out parts of the liver and spleen. The livers of these animals were gray, eroded, nasty looking.

Private Charlotte Godwin stood beside Captain Denny and handed him tools. She thought he looked nervous, jumpy, inside his space suit. He pulled a spleen out of a monkey. It was speckled with white spots, as hard as a rock, a biological bomb ticking with hot agent. After a while, he handed the scissors to her and gave her a chance to open up a monkey. It frightened her and gave her a big rush. She was doing a hot necropsy in Level 4, perhaps the most dangerous work in a space suit. This was a rocket ride, and it thrilled her. Her hands worked within a membrane's thickness of a death worse than any death in combat. She found herself racing to finish the job. She noticed that the monkey's eyes were open. It was as if the monkey was looking at her while she worked. She wanted to reach out and close the monkey's eyes. She thought, Is my face the last thing they see?

Inside

The day wore on, and people began using up their batteries. They could see that the daylight was starting to go, because some windows at the ends of the hallways were getting dark. Jerry Jaax made people rest every now and then. They sat on the floor with blank looks on their faces, exhausted, or they loaded syringes with drugs. Meanwhile, Jerry went from person to person, trying to gauge the level of exhaustion. "HOW ARE YOU DOING? ARE YOU TIRED? DO YOU WANT TO GO OUT?"

Nobody wanted to go out.

The team inside the building maintained radio contact with Gene Johnson outside the building. He had supplied them with hand-held short-wave radios that operated on a military band. He hadn't given them ordinary walkie-talkies because he didn't want anybody listening to the talk, especially the news media, who might make a tape recording of the chatter. It seemed less likely that anyone would listen to these radios.

Something went wrong with one of the soldiers' suits. She was a specialist named Rhonda Williams. Her blower cut off, and her suit began to go limp until it stuck to her sweaty scrub suit, and she felt contaminated air creeping around her. "MY AIR'S GOING OFF," she shouted. She kept working. She couldn't leave her post. Her battery was failing. She discovered that she did not have a spare battery on her belt. All the others had used their spare batteries.

When Rhonda announced that her air was shutting down, it caused a commotion. Jerry wanted to evacuate her from the building. He ran down the hall to the air-lock door, where a soldier was stationed with a short-wave radio. Jerry grabbed the radio and called Gene Johnson, shouting through his helmet, "WE'VE GOT A LADY WHO'S LOSING HER BATTERY."

Gene answered, "We need to get a battery and send it in with someone. Can you wait?"

"NO. SHE'S COMING OUT. SHE'S LOSING HER AIR," Jerry said.

Abruptly, the soldier by the door told Jerry that he had an extra battery. Jerry said over the radio: "WAIT—WE HAVE AN EXTRA ONE."

The soldier ran down the hallway to Rhonda, grinned at her, and said, "YOUR BATTERY IS HERE."

People started laughing. He clipped it to Rhonda's belt.

She thought, Oh, my God, they're going to unlock my old battery, it's going to stop my blowers. She said, "WAIT A MINUTE! MY AIR'S GOING TO GO OFF!"

"DON'T WORRY. IT'S JUST FOR A SECOND WHILE WE SWITCH YOU OVER," he said. Rhonda was rattled and was ready to leave. She was wondering if she had caught the virus during the moments when her air pressure had been lost. Jerry decided to send her out with Charlotte Godwin, who seemed to be getting tired. On the radio, he said to Gene, "I HAVE TWO COMING OUT."

On Gene's side, a near panic was occurring. A television van had just showed up. Gene was appalled. He didn't want the cameras to start rolling just as two women in space suits were extracted from the building. He said to Jerry, "We're jammed. We can't move them out. We've got TV cameras out here."

"I'M SENDING THEM OUT," Jerry said.

"All right. Send them out," Gene said. "We'll give the cameras a show."

Jerry pounded on the door of the gray area, and the decon man opened it. He was a sergeant. He wore a space suit. He held a pump sprayer filled with bleach, and a flashlight. Rhonda and Charlotte walked into the gray area, and the sergeant told them to hold their arms straight out at their sides. He played the flashlight over their space suits, checking for damage or leaks.

Rhonda noticed that he had a strange look on his face.

"YOU HAVE A HOLE IN YOUR SUIT," he said.

I *knew* this was going to happen, she thought.

"WHERE DID YOU GET IT?" he asked.

"I DON'T KNOW!"

He slapped a piece of tape over the hole. Then he washed the two soldiers down with bleach, spraying it all over them, and pounded on the door that led to the staging room. Someone opened it, and they went out. Immediately the support team opened their head bubbles and peeled off their suits. Their scrub suits, underneath, were soaked with sweat. They began to shiver.

"There's a television-news van out front," Gene said.

"I had a hole in my suit," Rhonda said to him. "Did I get the virus?"

"No. You had enough pressure in your suit to protect you the whole time." He hurried them outdoors. "Get into the van and lie down," he said. "If anybody asks you any questions, keep your mouths shut."

They couldn't find their clothes in the van. They rolled themselves up in some overcoats to keep warm and lay down on the seats, out of sight.

The television crew parked their van near the front door of

the monkey house, and the reporter began to poke around, followed by a cameraman. The reporter knocked on the front door and rang the buzzer—no answer. He peered in the front windows—the curtains were drawn, and he couldn't see anything. Well, nothing was happening in there. This place was deserted. He and the cameraman didn't notice the white vehicles parked behind the building, or if they noticed them, it didn't seem interesting. There was nothing going on here.

The television men returned to their van and sat in it for a while, hoping that something would happen or that someone would show up so that they could get some sound bites for the evening news, but this was getting to be boring, it was an awfully cold day, and the light was fading. It did not occur to them to go around to the side of the building and point their video camera toward a window. If they had done that, they would have gotten enough footage to fill the entire evening news, with something left over for CBS's *60 Minutes.* They would have gotten footage of soldiers in space suits smeared with Ebola blood, engaged in the first major biohazard mission the world ever knew, and they would have gotten shots of biohazard buddies coming out into the staging area in pairs and being stripped of their suits by the support teams. But the news crew didn't walk around the building, and so as far as I know, there is no video footage of the Reston action.

Meanwhile, the two women lay on their backs in the van for many minutes. Suddenly the television crew left. Gene Johnson, poking his head around the corner of the building, reported that the coast was clear. The women got dressed and then hurried off to relieve themselves in the wooded area behind the building. That was where they found the needles —two used hypodermic syringes with needles attached to them. The needles were uncapped and bare, obviously used. It was impossible to tell how long they had been lying in the grass. Some of the safety people put on gloves and picked up the needles, and as they searched the area, they found more needles in the grass.

The last person to come out was Jerry Jaax. He emerged around six in the evening, having lost between five and ten pounds of weight. It was fluid loss from sweating, and his face was ashen. His hair, instead of looking silver, looked white.

There was no food for the soldiers, and they were hungry and thirsty. The soldiers took a vote on where to eat, and it came out in favor of Taco Bell. Gene Johnson said to them, "Don't tell anybody why you are here. Don't answer any questions."

The caravan started up, engines roaring in the cold, and headed for Taco Bell. The soldiers ordered soft tacos with many jumbo Cokes to replace the sweat they'd lost inside their space suits. They also ordered a vast number of cinnamon twists—everything to go—yeah, put it in boxes, and hurry, please. The employees were staring at them. The soldiers looked like soldiers, even in jeans and sweat shirts— the men were bulked up and hard-looking, with crew cuts and metal-framed military eyeglasses and a few zits from too much Army food, and the women looked as if they could do fifty push-ups and break down a weapon. A man came up to Sergeant Klages while he was waiting for his food and said, "What were you doing over there? I saw all those vans." Sergeant Klages turned his back on the man without saying a word.

After midnight on the water bed in the master bedroom of the Jaax house on the slopes of Catoctin Mountain, Nancy and Jerry Jaax caught up on the news while their daughter, Jaime, slept beside them. Jerry told her that the day's operation had gone reasonably well and that no one had stuck himself or herself with a needle. He told her he hadn't realized how lonely it is inside a biohazard suit.

Nancy wrapped herself around him and rested her head against his neck in the way they had held each other since college. She thought he looked shrunken and thin. He was physically more exhausted than she had seen him in years. She picked up Jaime and carried her to bed, then returned and folded herself around her husband. They fell asleep holding each other.

A Bad Day

For the past several days and nights, an Army scientist named Thomas Ksiazek had been working in his space suit in a Level 4 lab trying to develop a rapid test for Ebola virus in blood and tissue. He got the test to work. It was called a rapid Elisa test, and it was sensitive and easy to perform. He tested urine and blood samples from Milton Frantig, the man who had vomited on the lawn and who was now in an isolation room at Fairfax Hospital. Frantig came up clean. His urine and blood did not react to the Ebola test. It looked as if he had the flu. This was a mystery. Why weren't these guys breaking with Ebola?

The weather warmed up and turned sunny, and the wind shifted around until it blew from the south. On the second day of the massive nuking—Wednesday—the Army caravan flowed with commuter traffic to Reston and deployed behind the monkey house. Things went more smoothly. By eight o'clock in the morning, the teams had begun their insertions. Gene Johnson brought a floodlight, and they set it up in the gray corridor.

Jerry Jaax went in first and fed the monkeys. He made his rounds with Sergeant Amen, checking each room, and here and there they found monkeys dead or in terminal shock. In a lounge, they found some chairs, and dragged them into a hallway and arranged them in a semicircle so that the soldiers could sit on them while they took their rest breaks and filled up syringes. As the day wore on, you could see exhausted soldiers and civilians in orange space suits, men and women, their head bubbles clouded with condensation, sitting on the chairs in the hallway, loading syringes with T-61 and sorting boxes full of blood tubes. Some talked with each other by shouting, and others just stared at the walls.

At midmorning, Jerry Jaax was working in Room C. He decided to take a break to rest and check up on his people. He left the room in charge of Sergeants Amen and Klages while he went out into the hallway. Suddenly there was a commotion in Room C, and the monkeys in that room burst out in wild screeches. Jerry ran back to the room, where he found the sergeants outside the door, looking in, in a state of alarm.

"WHAT HAPPENED?"

"A MONKEY ESCAPED, SIR."

"AW, SHIT!" Jaax roared.

The animal had bolted past Sergeant Amen as he opened the cage, and the sergeants had immediately run out of the room and shut the door behind them.

A loose monkey—this was what Jerry had feared the most. They can leap long distances. He had been bitten by monkeys himself, and he knew what that felt like. Those teeth went in deep.

They looked into the room through the window in the door. The whole room had exploded in activity, monkeys whirling in their cages and shaking them violently, giving off high, excited whoops. There were about a hundred screaming monkeys in that room. But where was the loose monkey? They couldn't see it.

They found a catching net, a pole with a baglike net at the end. They opened the door and edged into the room.

The events that followed have a dreamlike quality in people's memories, and the memories are contradictory. Specialist Rhonda Williams has a memory that the monkey escaped from the room. She says she was sitting on a chair when it happened, that she heard a lot of shouting and suddenly the animal appeared and ran under her feet. She froze in terror, and then burst out laughing—nervous, near-hysterical laughter. The animal was a small, determined male, and he was not going to let these humans get near him with a net.

Jerry Jaax insists that the monkey never got out of the room. It is possible that the monkey ran under Specialist Williams's feet and then was chased back into the room again.

The loose monkey was very frightened and the soldiers were very frightened. He stayed in the room for a while, running back and forth across the cages. The other monkeys apparently grew angry at this and bit at the monkey's toes. The monkey's feet began to bleed, and pretty soon it had tracked blood all over the room. Jerry got on the radio and reported that a monkey was loose and bleeding. Gene Johnson told him to do whatever had to be done. How about shooting the monkey? Bring in a handgun, like an Army .45. Jerry didn't like that idea. Looking into the room, he noticed that the loose monkey was spending most of its time hiding behind the cages. If you tried to shoot the monkey, you'd be firing into the cages, and the bullet could hit a cage or a wall and might ricochet inside the room. Getting a gunshot wound was bad enough under any circumstances, but even a mild wound in this building might be fatal. He decided that the safest procedure would be to go into the room and capture the monkey with the net. He took Sergeant Amen with him.

As they entered the room, they could not see the monkey. Jerry proceeded forward slowly, holding the net up, ready to swipe it at the monkey. But where was it? He could not see very well. His faceplate was covered with sweat, and the light was dim in the room. He might as well have been swimming underwater. He edged slowly forward, keeping his body away from the cages on either side, which were filled with hysterical, screaming, leaping, bar-rattling monkeys. The sound of monkeys raising hell was deafening. He was afraid of being bitten by a monkey if he came too close to a cage. So he stayed in the middle of the room as he went forward, while Sergeant Amen followed him, holding a syringe full of drugs on a pole.

"BE CAREFUL, SERGEANT," he said. "DON'T GET BITTEN. STAY BACK FROM THE CAGES."

He edged his way from cage to cage, looking into each one, trying to see through it toward the shadowy wall behind. Suddenly he saw a flicker of movement out of the corner of his eye, and he turned with the net, and the monkey went soaring through the air *over* him, making a twelve-foot jump from one side of the room to the other.

"GET HIM! HE'S OVER HERE!" he said. He waved the net, slammed it around over the cages, but the monkey was gone.

He walked through the room again, slowly. The monkey flung itself across the room, a huge, tail-swinging leap. This animal was airborne whenever it moved. Jerry waved his net and missed. "SON OF A BITCH!" he shouted. The monkey was too fast for him. He would spend ten or fifteen minutes searching the room, squinting past the cages. If he found the monkey, the monkey would leap to the other side of the room. It was a small monkey, built for life in the trees. He thought, This environment favors the monkey over us. We don't have the tools to handle this situation. We are not in control here—we are along for the ride.

Outside the building, Colonel C. J. Peters stopped by to observe the operation. He was dressed in Levi's and a sweater, along with sandals and socks, even though it was a cold day. With his sandals and mustache, he appeared to be a sixties type or some sort of a low-grade employee, maybe a

258 THE HOT ZONE

janitor. He noticed a stranger hanging around the front of the building. Who was this? Then the man started to come around the side of the building. He was obviously after something, and he was getting too close to the action. C. J. hurried forward and stopped the man and asked him what he was doing.

He said he was a reporter from *The Washington Post*. "What's happening around here?" he asked C. J.

"Well—aw—nothing much is happening," C. J. replied. He was suddenly very glad he had not worn his colonel's uniform today—for once, his bad habits had paid off. He did not encourage the reporter to come around to the side of the building and have a look in through the window. The reporter left shortly afterward, having seen and heard nothing of interest. *The Washington Post* suspected that something funny was happening at the monkey house, but the reporters and editors who worked on the story couldn't quite get to the bottom of it.

"This monkey knows nets," Jerry shouted to the sergeant. The monkey was not going to let himself be caught by some fool of a human wearing a plastic bag. They decided to leave him in the room overnight.

Meanwhile, the surviving monkeys were becoming increasingly agitated. The teams killed most of the monkeys this day, working straight through until after dark. Some of the soldiers began to complain that they were not being given enough responsibility, and so Jerry let them take over more of the hazardous work from the officers. He assigned Specialist Rhonda Williams to duty at the euthanasia table with Major Nate Powell. The major laid a drugged monkey on the table, holding its arms behind its back in case it woke up, while Rhonda uncapped a syringe and gave the monkey a heart stick—plunged the needle into the chest between the ribs, aiming for the heart. She pushed the plunger, sending a load of drugs into the heart, which killed the monkey instantly. She pulled the needle out, and a lot of blood squirted out of the puncture wound. That was a good sign; it meant she had punctured the heart. If she got blood on her gloves, she rinsed them in a pan of bleach, and if she got blood on her space suit, she wiped it down with a sponge soaked in bleach.

It was awful when she missed the heart. She pushed the plunger, the poison flooded the animal's chest around the heart, and the monkey jumped. It doubled up, its eyes moved, and it seemed to struggle. This was only a death reflex, but she gasped, and her own heart jumped.

Then Colonel Jaax put her to work at the bleed table with Captain Haines, and presently she began drawing blood from unconscious monkeys. She inserted a needle into the animals' leg vein and drew the blood. Their eyes were open. She didn't like that. She felt they were staring at her.

She was bleeding a monkey when suddenly she thought its eyes moved, and it seemed to be trying to sit up. It was awake. It looked at her in a daze and reached out and grabbed her by the hand, the one that was holding the syringe. The monkey was very strong. The needle came out of its thigh, and blood spurted out. Then the animal started pulling her hand toward its mouth! It was trying to bite her hand! She screamed: "GRAB HIM, SOMEBODY, PLEASE! HE'S GETTING UP!" Captain Haines caught the monkey's arms and pinned it to the table, shouting, "WE HAVE ONE THAT'S AWAKE! NEED KETAMINE!"

The needle, in coming out of the monkey, had cut the monkey's leg vein. Immediately a ball of blood the size of a baseball formed in the monkey's leg. It just got bigger and bigger, the blood pouring under the skin, and Rhonda almost burst into tears. She pressed her hands on the blood ball to stop the internal bleeding. Through her gloves, she could feel the blood swelling. A ball of Ebola blood.

A soldier hurried over and hit the monkey with a double load of ketamine, and the monkey went limp.

During the crisis, Peter Jahrling spent every day wearing a space suit in his lab, running tests on monkey samples, trying to determine where and how the virus was spreading, and trying to get a pure sample of the virus isolated. Meanwhile, Tom Geisbert pulled all-nighters, staring at the cellscapes through his microscope.

Occasionally they met each other in an office, and closed the door.

"How are you feeling?"

"Tired, but otherwise I'm okay."

"No headaches?"

"Nope. How are you feeling?"

"Fine."

They were the discoverers of the strain, and it seemed that they would have the chance to give it a name, provided they could isolate it, and provided it didn't isolate them first.

Jahrling went home for dinner with his family, but after he had read his children their stories and put them to bed, he returned to the Institute and worked until late. The whole Institute was lit up with activity, all the hot labs full of people and operating around the clock. Soon he had stripped nude in the locker room, and he was putting on his scrubs, and then he was wearing his space suit, feeling sleepy, warm, and full of dinner, as he faced the steel door blazed with the red flower, reluctant to take another step forward. He opened the door and went through to the hot side.

He had been testing his and Geisbert's blood all along, and he wondered if the virus would suddenly show up in it. He didn't think it was likely. I didn't stick the flask close to my nose, I kind of just waved my hand over it. They used to do that all the time in hospital labs with bacteria. It used to be standard procedure to sniff cultures in a lab—that was how you learned what bacteria smelled like, how you learned that some kinds smell like Welch's grape juice.

The question of whether he, Peter Jahrling, was infected with Ebola had become somewhat more pressing since the animal caretaker had puked on the lawn. That guy had not cut himself or stuck himself with a needle. Therefore, if that guy was breaking with Ebola, he might have caught it by breathing it in the air.

Jahrling carried some slides containing spots of his own blood serum into his closet, shut the door, turned out the light. He let his eyes adjust to the darkness, and had the usual struggle to see anything in the microscope through his faceplate. Then the panorama swam into view. It was the ocean of his blood, stretching in all directions, grainy and mysterious, faintly glowing with green. This was a normal glow, nothing to get excited about, that faint green. If the green brightened into a hotter glow, that would signify that his blood was inhabited by Ebola. And what if his blood glowed? How would he judge if it was really glowing? How green is green? How much do I trust my tools and my perceptions? And if I'm convinced my blood is glowing, how am I going to report the results? I'll need to tell C. J. Maybe I won't have to go into the Slammer. I could be biocontained right here in my own lab. I'm in Biosafety Level 4 right now. I'm already in isolation. Who can I infect here in my lab? Nobody. I could live and work in here if I go positive for Ebola.

Nothing glowed. Nothing reacted to his blood. His blood was normal. Same with Tom Geisbert's blood. As to whether their blood would glow tomorrow or the next day or the day after that, only time would tell, but he and Geisbert were climbing out of the incubation period.

At eleven o'clock at night, he decided it was time to go home, and he entered the air lock and pulled the chain to start the decon cycle. He was standing in gray light in the gray zone, alone with his thoughts. He couldn't see much of anything in here, in the chemical mist. He had to wait seven minutes for the cycle to complete itself. His legs were killing him. He was so tired he couldn't stand up. He reached up with his hands and grabbed the pipes that fed chemicals into the shower, to hold himself up. The warm liquid ran over his space suit. He felt comfortable and safe in here, surrounded by the sloshing noises of virus-killing liquids and the hiss of air and the ruffling sensation across his back as the chemicals played over his suit. He fell asleep.

He jerked awake when the final blast of water jets hit him, and he found himself slumped against the wall of the air lock, his hands still gripped around the pipes. If it hadn't been for that last jet of water, he would not have woken up. He would have slid down the wall and curled up in the corner of the air lock, and probably would have stayed there all night, sound asleep, while the cool, sterile air flowed through his suit and bathed his body, nude inside its cocoon, at the heart of the Institute.

Specialist Rhonda Williams was standing in the main corridor of the monkey house, afraid she would end up in the Slammer. There was no sound except the roar of air in her helmet. The corridor stretched in both directions to infinity, strewn with cardboard boxes and trash and monkey biscuits. Where were the officers? Where was Colonel Jaax? Where was everybody? She saw the doors leading to the monkey rooms. Maybe the officers were in there.

Something was coming down the corridor. It was the loose monkey. He was running toward her. His eyes were staring at her. Something glittered in his hand—he was holding a syringe. He waved it at her with a gesture that conveyed passionate revenge. He wanted to give her an injection. The syringe was hot with an unknown agent. She started to run. Her space suit slowed her down. She kept running, but the hallway stretched on forever, and she couldn't reach the end. Where was the door out of here? There was no door! There was no way out! The monkey bounded toward her, its terrible eyes fixed on her—and the needle flashed and went into her suit. . . . She woke up in her barracks room.

Decon

DECEMBER 7, THURSDAY

Nancy Jaax awoke at four o'clock in the morning to the sound of the telephone ringing. It was her brother, calling from a pay telephone at the hospital in Wichita. He said that their father was dying. "He's very, very bad, and he's not going to make it," he said. Their father was in cardiac failure, and the doctor had been asking if the family wanted him to undertake extreme lifesaving measures. Nancy thought only briefly about this and told her brother not to do it. Her father was down to ninety pounds, just skin and bones, and he was in pain and miserable.

She woke up Jerry and told him that her father would probably die today. She knew she would have to go home, but should she try to fly home today? She could arrive in Wichita by afternoon, and he might still be alive. She might be able to have a last farewell with him. She decided not to fly home. She felt that she couldn't leave her job in the middle of the Reston crisis, that it would be a dereliction of her post.

The telephone rang again. It was Nancy's father calling from his hospital room. "Are you coming home, Nancy?" he asked. He sounded wheezy and faint.

"I just can't get away right now, Dad. It's my work. I'm in the middle of a serious outbreak of disease."

"I understand," he said.

"I'll see you at Christmas, Dad."

"I don't think I'll make it that long, but, well, you never know."

"I'm sure you will make it."

"I love you, Nancy."

"I love you, too."

In the blackness before dawn, she and Jerry got dressed, she in her uniform, he in civilian clothes, and he headed off for the monkey house. Nancy stayed at home until after the children had woken up, and she fixed them some cereal. She sent the children off on the school bus and drove to work. She went to Colonel C. J. Peters and told him that her father was probably going to die today.

"Go home, Nancy," he said.

"I'm not going to do that," she replied.

The dead monkeys began coming in after lunch. A truck would bring them twice a day from Reston, and the first shipment would end up in Nancy's air lock while she was suiting up. Usually there would be ten or twelve monkeys in hatboxes.

The rest of the monkeys that came out of the monkey house—the vast majority of them, two or three tons' worth —were placed in triple biohazard bags, and the bags were decontaminated, taken out of the building, and placed in steel garbage cans. Hazleton employees then drove them to an incinerator owned by the company, where the monkeys were burned at a high temperature, high enough to guarantee the destruction of Ebola organisms.

Some of the monkeys had to be examined, however, to see if and where the virus was spreading inside the building. Nancy would carry the hatboxes into suite AA-5 and work on the monkeys until after midnight with her partner and a civilian assistant. They hardly spoke to one another, except to point to a tool or to a sign of disease in a monkey.

Thoughts about her father and her childhood came to Nancy that day. Years earlier, as a girl, she had helped him during plowing season, driving his tractor from afternoon until late at night. Moving at a pace not much faster than a mule, it plowed furrows along a strip of land a half a mile long. She wore cutoff shorts and sandals. It was loud and hot on the tractor, and in the emptiness of Kansas she thought about nothing, drowned in the roar of the engine as the sun edged down to the horizon and the land grew dark and the moon appeared and climbed high. At ten o'clock her father would take over and plow for the rest of the night, and she would go to bed. At sunrise, he would wake her up, and she would get back on the tractor and keep on plowing.

"SPONGE," she mouthed to her buddy.

He mopped up some blood from the monkey, and Nancy rinsed her gloves in the pan of green EnviroChem.

Her father died that day, while Nancy worked in the hot suite. She flew home to Kansas and arrived by taxi on Saturday morning at her family's plot at a graveyard in Wichita just as the funeral service began. It was a cold, rainy day, and a tiny knot of people holding umbrellas huddled around a preacher by a stone wall and a hole in the earth. Lieutenant Colonel Nancy Jaax moved forward to see more clearly, and her eyes rested on something that she had not quite anticipated. It was a flag draped over the casket. He had been a veteran, after all. The sight broke her down, and she burst into tears.

At four o'clock in the afternoon, Thursday, December 7, the last monkey was killed and bagged, and people began deconning out. They had had a bad time trying to catch the little monkey that had escaped; it took hours. Jerry Jaax had entered the room where it was hiding and spent two or three hours chasing it in circles with a net. Finally the monkey got itself jammed down in a crack behind a cage with its tail sticking out, and Sergeant Amen hit the tail with a massive dose of anesthetic. In about fifteen minutes, the monkey became still, and they dragged it out, and it went the way of the other monkeys, carried along in the flow of material.

They radioed Gene Johnson to tell him that the last monkey was dead. He told Sergeant Klages to explore the building, to make sure that there were no more live monkeys in any rooms. Klages discovered a chest freezer in a storage room. It looked sinister, and he radioed to Johnson: "GENE, I'VE GOT A FREEZER HERE."

"Check it out," Johnson replied.

Sergeant Klages lifted the lid. He found himself staring into the eyes of frozen monkeys. They were sitting in clear plastic bags. Their bodies streamed with blood icicles. They were monkeys from Room F, the original hot spot of the outbreak, some of monkeys that had been sacrificed by Dan Dalgard. He shut the lid and called Johnson on the radio:

"GENE, YOU'RE NOT GOING TO BELIEVE WHAT I'VE FOUND IN THIS FREEZER. I'VE GOT TEN OR FIFTEEN MONKEYS."

"Aw, shit, Klages!"

"WHAT SHOULD I DO WITH THEM?"

"I don't want any more problems with monkeys! No more samples! Decon them!"

"I ALSO FOUND SOME VIALS OF SEDATIVE."

"Decon it, baby! You don't know if any dirty needles have been stuck in those bottles. Everything comes out of this building! Everything comes out!"

Sergeant Klages and a civilian, Merhl Gibson, dragged the bags out of the freezer. They tried to cram the monkeys into hatboxes, but they didn't fit. They were twisted into bizarre shapes. They left them in the hallway to thaw. The decon teams would deal with them tomorrow.

The 91-Tangos shuffled out through the air-lock corridor, two by two, numb and tired beyond feeling, soaked with sweat and continual fear. They had collected a total of thirty-five hundred clinical samples. They didn't want to talk about the operation with each other or with their officers.

When the team members left for Fort Detrick, they noticed that Gene Johnson was sitting on the grass under the tree in front of the building. He didn't want to talk to anyone, and they were afraid to talk to him. He looked terrible. His mind was a million miles away, in the devastated zone inside the building. He kept going over and over what the kids had done. If the guy has the needle in his right hand, you stand on his left. You pin the monkey's arms behind so it can't turn around and bite you. Did anyone cut a finger? So far, it looked as if all the kids had made it.

The decon team suited up immediately while the soldiers were coming out of the building. It was now after dark, but Gene Johnson feared Ebola so much that he did not want to let the building sit untouched overnight.

The decon team was led by Merhl Gibson. He put on a space suit and explored the building to get a sense of what needed to be done. The rooms and halls were bloodstained and strewn with medical packaging. Monkey biscuits lay everywhere and crunched underfoot. Monkey feces lay in loops on the floor and was squiggled in lines across the walls and printed in the shapes of small hands. He had a brush and a bucket of bleach, and he tried to scrub a wall.

Then he called Gene on the radio. "GENE, THE SHIT IN HERE IS LIKE CEMENT, IT WON'T COME OFF."

"You do what's best. Our orders are to clean this place up."

"WE'LL TRY TO CHIP IT OFF," Gibson said.

The next day, they went to a hardware store and bought putty knives and steel spatulas, and the decon team went to work chipping the walls and floor. They almost suffocated from the heat inside their suits.

Milton Frantig, the man who had thrown up on the lawn, had now been kept in isolation at Fairfax Hospital for several days. He was feeling much better, his fever had vanished, he had not developed any nosebleeds, and he was getting restless. Apparently he did not have Ebola. At any rate, it did not show up in his blood tests. Apparently he had a mild case of flu. The C.D.C. eventually told him he could go home.

By day nineteen after the whiffing incident, when they hadn't had any bloody noses, Peter Jahrling and Tom Geisbert began to regard themselves as definite survivors. The fact that Dan Dalgard and the monkey workers had so far shown no signs of breaking with Ebola also reassured them, although it was very puzzling. What on earth was going on with this virus? It killed monkeys like flies, they were dripping virus from every pore, yet no human being had crashed. If the virus wasn't Ebola Zaire, what was it? And where had it come from? Jahrling believed that it must have come from Africa. After all, Nurse Mayinga's blood reacted to it. Therefore, it must be closely related to Ebola Zaire. It was behaving like the fictional Andromeda strain. Just when we thought the world was coming to an end, the virus slipped away, and we survived.

The Centers for Disease Control focused its efforts on trying to trace the source of the virus, and the trail eventually led back to the Ferlite Farms monkey-storage facility near Manila. All of the Reston monkeys had come from there. The place was a way station on their trip from the forests of Mindanao to Washington. Investigators found that monkeys had been dying in large numbers there, too. But it looked as though no Philippine monkey workers had become sick either. If it was an African virus, what was it doing in the Philippines? And why weren't monkey handlers dying? Yet the virus was able to destroy a monkey. Something very strange was going on here. Nature had seemed to be closing in on us for a kill, when she suddenly turned her face away and smiled. It was a Mona Lisa smile, the meaning of which no one could figure out.

DECEMBER 18, MONDAY

The decon team scrubbed the building with bleach until they took the paint off the concrete floors, and still they kept scrubbing. When they were satisfied that all of the building's inside surfaces had been scoured, they moved on to the final stage, the gas. The decon team taped the exterior doors, windows, and vents of the building with silver duct tape. They taped sheets of plastic over the exterior openings of the ventilation system. They made the building airtight. At various places inside the monkey house, they set out patches of paper saturated with spores of a harmless bacterium known as *Bacillus subtilis niger*. These spores are hard to kill. It is believed that a decon job that kills *niger* will kill almost anything.

The decon team brought thirty-nine Sunbeam electric frying pans to the monkey house. Sunbeam electric frying pans are the Army's tool of choice for a decon job. The team laid an electric cable along the floor throughout the building, strung with outlets, like a cord for Christmas-tree lights. At points along the cable, they plugged in the Sunbeam frying pans. They wired the cable to a master switch. Into each Sunbeam frying pan they dropped a handful of disinfecting crystals. The crystals were white and resembled salt. They dialed the pans to high. At 1800 hours on December 18, someone threw the master switch, and the Sunbeams began to cook. The crystals boiled away, releasing formaldehyde gas. Since the building's doors, windows, and vents were taped shut, the gas had nowhere to go, and it stayed inside the building for three days. The gas penetrated the air ducts, soaked the offices, got into drawers in the desks, and got inside pencil sharpeners in the drawers. It infiltrated Xerox machines and worked its way inside personal computers and inside the cushions of chairs and fingered down into the floor drains until it touched pools of lingering bleach in the water traps. Finally the decon team, still wearing space suits, went back inside the building and collected the spore samples. The Sunbeam treatment had killed the *niger*.

There is an old piece of wisdom in biohazard work that goes like this: you can never know when life is exterminated. Life will survive almost any blitz. Total, unequivocal sterilization is extremely difficult to achieve in practice and is almost impossible to verify afterward. However, a Sunbeam cookout that lasts for three days and exterminates all samples of *niger* implies success. The monkey house had been sterilized. Ebola had met opposition. For a short while, until life could re-establish itself there, the Reston Primate Quarantine Unit was the only building in the world where nothing lived, nothing at all.

The Most Dangerous Strain

1990 JANUARY

The strain of Ebola virus that had erupted near Washington went into hiding somewhere in the rain forest. The cycling went on. The cycling must always go on if the virus is to maintain its existence. The Army, having certified that the monkey house had been nuked, returned it to the possession of Hazleton Research Products. Hazleton began buying more monkeys from the Philippines, from the same monkey house near Manila, and restocked the building with crab-eating monkeys that had been trapped in the rain forests of Mindanao. Less than a month later, in the middle of January, some of the monkeys in Room C began dying with bloody noses. Dan Dalgard called Peter Jahrling. "Looks like we're affected again," he said.

The virus was Ebola. It had come from the Philippines. This time, since there had been no human casualties during the first outbreak, the Army, the C.D.C., and Hazleton jointly decided to isolate the monkeys—leave them alone and let the virus burn. Dan Dalgard hoped to save at least some of the monkeys, and his company did not want the Army to come back with space suits.

What happened in that building was a kind of experiment.

Now they would see what Ebola could do naturally in a population of monkeys living in a confined air space, in a kind of city, as it were. The Ebola Reston virus jumped quickly from room to room, and as it blossomed in the monkeys, it seemed to mutate spontaneously into something that looked quite a lot like influenza. But it was an Ebola flu. The monkeys died with great quantities of clear mucus and green mucus running from their noses, mixed with blood that would not clot. Their lungs were destroyed, rotten and swimming with Ebola virus. They had pneumonia. When a single animal with a nosebleed showed up in a room, generally 80 percent of the animals died in that room shortly afterward. The virus was extraordinarily contagious in monkeys. The Institute scientists suspected that they were seeing a mutant strain of Ebola, something new and a little different from what they had seen just a month before, in December, when the Army had nuked the monkey house. It was frightening— it was as if Ebola could change its character fast—and could look like the flu. As if a different strain could appear in a month's time. The clinical symptoms of the disease served as a reminder of the fact that Ebola is related to certain kinds of flu-like illnesses seen in human children. It seemed that the virus could adapt quickly to new hosts, and that it could change its character spontaneously and rapidly as it entered a new population.

Ebola apparently drifted through the building's air-handling ducts. By January 24, it had entered Room B, and monkeys in that room started going into shock and dying with runny noses, red eyes, and masklike expressions on their faces. In the following weeks, the infection entered Rooms I, F, E, and D, and the animals in these rooms virtually all died. Then, in mid-February, a Hazleton animal caretaker who will be called John Coleus was performing a necropsy on a dead monkey when he cut his thumb with a scalpel. He had been slicing apart the liver, one of the favorite nesting sites of Ebola. The scalpel blade, smeared with liver cells and blood, went deep into his thumb. He had had a major exposure to Ebola.

The liver that he had been cutting was rushed to USAMRIID for analysis. Tom Geisbert looked at a piece of it under his microscope and, to his dismay, found that it was "incredibly hot—I mean, wall to wall with virus." Everyone at the Institute thought John Coleus was going to die. "Around here," Peter Jahrling told me, "we were frankly fearful that this guy had bought the farm." The C.D.C. decided not to put him into isolation. So Coleus visited bars and drank beer with his friends. While he was incubating the virus.

"Here at the Institute," Peter Jahrling said, "we were absolutely appalled when that guy went out to bars, drinking. Clearly the C.D.C. should not have let that happen. This was a serious virus and a serious situation. We don't know a whole lot about the virus. It could be like the common cold —it could have a latency period when you are shedding virus before you develop symptoms—and by the time you know you are sick, you might have infected sixteen people. There's an awful lot we don't know about this virus. We don't know where it came from, and we don't know what form it will take when it appears next time."

John Coleus had a minor medical condition that required surgery. Doctors performed the operation while he was in the incubation period after his exposure to Ebola. There is no record indicating that he bled excessively during the surgery. He came through fine, and he is alive today, with no ill effects from his exposure.

As for the monkey house, the entire building died. The Army didn't have to nuke it. It was nuked by the Ebola Reston virus. Once again, there were no human casualties. However, something eerie and perhaps sinister occurred. A total of four men had worked as caretakers in the monkey house:

Jarvis Purdy, who had had a heart attack; Milton Frantig, who had thrown up on the lawn; John Coleus, who had cut his thumb; and a fourth man. All four men eventually tested positive for Ebola Reston virus. They had all been infected with the agent. The virus had entered their bloodstreams and multiplied in their cells. Ebola proliferated in their bodies. It cycled in them. It carried on its life inside the monkey workers. But it did not make them sick, even while it multiplied inside them. If they had headaches or felt ill, none of them could recall it. Eventually the virus cleared from their systems naturally, disappeared from their blood, and as of this writing none of the men was affected by it. They are among the very, very few known human survivors of Ebola virus. John Coleus certainly caught the virus when he cut himself with a bloody scalpel, no question about that. What is more worrisome is that the others did *not* cut themselves, yet the virus still entered their bloodstreams. It got there somehow. Most likely it entered their blood through contact with the lungs. It infected them through the air. When it became apparent to the Army researchers that three of the four men who became infected had not cut themselves, just about everyone at USAMRIID concluded that Ebola can spread through the air.

Dr. Philip Russell—the general who made the decision to send in the Army to stop the virus—recently said to me that although he had been "scared to death" about Ebola at the time, it wasn't until afterward, when he understood that the virus was spreading in the air among the monkeys, that the true potential for disaster sank in for him. "I was more frightened in retrospect," he said. "When I saw the respiratory evidence coming from those monkeys, I said to myself, My God, with certain kinds of small changes, this virus could become one that travels in rapid respiratory transmission through *humans*. I'm talking about the Black Death. Imagine a virus with the infectiousness of influenza and the

mortality rate of the black plague in the Middle Ages—
that's what we're talking about.''

The workers at Reston had had symptomless Ebola virus.
Why didn't it kill them? To this day, no one knows the
answer to that question. Symptomless Ebola—the men had
been infected with something like an Ebola cold. A tiny
difference in the virus's genetic code, probably resulting in a
small structural change in the shape of one of the seven
mysterious proteins in the virus particle, had apparently
changed its effects tremendously in humans, rendering it
mild or harmless even though it had destroyed the monkeys.
This strain of Ebola knew the difference between a monkey
and a person. And if it should mutate in some other direction . . .

One day in spring, I went to visit Colonel Nancy Jaax, to
interview her about her work during the Reston event. We
talked in her office. She wore a black military sweater with
silver eagles on the shoulder boards—she had recently made
full colonel. A baby parrot slept in a box in the corner. The
parrot woke up and squeaked.

"Are you hungry?" she asked it. "Yeah, yeah, I know."
She pulled a turkey baster out of a bag and loaded it with
parrot mush. She stuck the baster into the parrot's beak and
squeezed the baster bulb, and the parrot closed its eyes with
satisfaction.

She waved her hand at some filing cabinets. "Want to
look at some Ebola? Take your pick.''

"You show me," I said.

She searched through a cabinet and removed a handful of
glass slides, and carried them into another room, where a
microscope sat on a table. It had two sets of eyepieces so that
two people could look into it at the same time.

I sat down and stared in the microscope, into white noth-
ingness.

"Okay, here's a good one," she said, and placed a slide under the lens.

I saw a field of cells. Here and there, pockets of cells had burst and liquefied.

"That's male reproductive tissue," she said. "It's heavily infected. This is Ebola Zaire in a monkey that was exposed through the lungs in 1986, in the study that Gene Johnson and I did."

Looking at the slice of monkey testicle, I got an unpleasant sensation. "You mean, it got into the monkey's lungs and then moved to its testicles?"

"Yeah. It's pretty yucky," she said. "Now I'm going to make you dizzy. I'm going to show you the lung."

The scene shifted, and we were looking at rotted pink Belgian lace.

"This is a slice of lung tissue. A monkey that was exposed through the lungs. See how the virus bubbles up in the lung? It's Ebola Zaire."

I could see individual cells, and some of them were swollen with dark specks.

"We'll go to higher magnification."

The cells got bigger. The dark specks became angular, shadowy blobs. The blobs were bursting out of the cells, like something hatching.

"Those are big, fat bricks," she said.

They were Ebola crystalloids bursting out of the lungs. The lungs were popping Ebola directly into the air. My scalp crawled, and I felt suddenly like a civilian who had seen something that maybe civilians should not see.

"These lungs are very hot," she said in a matter-of-fact voice. "You see those bricks budding directly into the air spaces of the lung? When you cough, this stuff comes up your throat in your sputum. That's why you don't want someone who has Ebola coughing in your face."

"My God, it knows all about lungs, doesn't it?"

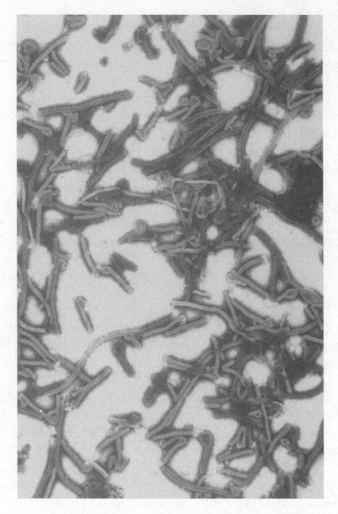

Ebola Zaire virus particles magnified 17,000 times. Note the loops at the ends of some particles, the so-called shepherd's crooks or eyebolts, which are typical of Ebola Zaire and its sisters. Photograph by Thomas W. Geisbert, USAMRIID.

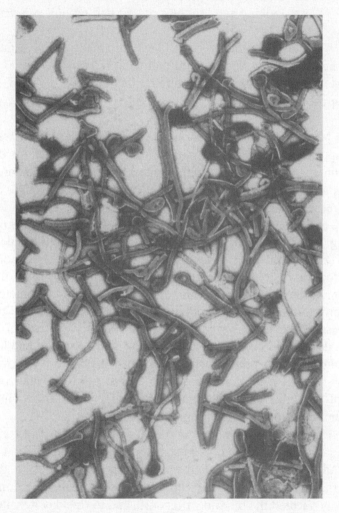

Ebola Reston virus particles. "The point is, you can't easily tell the difference between the two strains by looking." —Peter Jahrling. Photograph by Thomas W. Geisbert, USAMRIID.

"Maybe not. It might live in an insect, and insects don't have lungs. But you see here how Ebola has *adapted* to this lung. It's budding out of the lung, right straight into the air."

"We're looking at a highly sophisticated organism, aren't we?"

"You are absolutely right. This hummer has an established life cycle. You get into that what-if? game. What if it got into human lungs? If it mutates, it could be a problem. A *big* problem."

In March 1990, while the second outbreak at Reston was happening, the C.D.C. slapped a heavy set of restrictions on monkey importers, tightening the testing and quarantine procedures. The C.D.C. also temporarily revoked the licenses of three companies, Hazleton Research Products, the Charles River Primates Corporation, and Worldwide Primates, charging these companies with violations of quarantine rules. (Their licenses were later reinstated.) The C.D.C.'s actions effectively stopped the importation of monkeys into the United States for several months. The total loss to Hazleton ran into the millions of dollars. Monkeys are worth money. Despite the C.D.C.'s action against Hazleton, scientists at USAMRIID, and even some at the C.D.C., gave Dalgard and his company high praise for making the decision to hand over the monkey facility to the Army. "It was hard for Hazleton, but they did the right thing," Peter Jahrling said to me, summing up the general opinion of the experts.

Hazleton had been renting the monkey house from a commercial landlord. Not surprisingly, relations between the landlord and Hazleton did not flourish happily during the Army operation and the second Ebola outbreak. The company vacated the building afterward, and to this day it stands empty.

Peter Jahrling, a whiffer of Ebola who lived to tell about it, is now the principal scientist at USAMRIID. He and Tom Geisbert, following tradition in the naming of new viruses, named the strain they had discovered Reston, after the place where it was first noticed. In conversation, they sometimes refer to it casually as Ebola Reston. One day in his office, Jahrling showed me a photograph of some Ebola-virus particles. They resembled noodles that had been cooked al dente. "Look at this honker. Look at this long sucker here," Jahrling said, his finger tracing a loop. "It's Reston—oh, I was about to say it's Reston, but it isn't—it's Zaire. The point is, you can't easily tell the difference between the two strains by looking. It brings you back to a philosophical question: Why is the Zaire stuff hot for humans? Why isn't Reston hot for humans, when the strains are so close to each other? The Ebola Reston virus is almost certainly transmitted by some airborne route. Those Hazleton workers who had the virus—I'm pretty sure they got it through the air."

"Did we dodge a bullet?"

"I don't think we did," Jahrling said. "The bullet hit us. We were just lucky that the bullet we took was a rubber bullet from a twenty-two rather than a dumdum bullet from a forty-five. My concern is that people are saying, 'Whew, we dodged a bullet.' And the next time they see Ebola in a microscope, they'll say, 'Aw, it's just Reston,' and they'll take it outside a containment facility. And we'll get whacked in the forehead when the stuff turns out not to be Reston but its big sister."

C. J. Peters eventually left the Army to become the chief of the Special Pathogens Branch at the Centers for Disease Control. Looking back on the Reston event, he said to me one day that he was pretty sure Ebola had spread through the air. "I think the pattern of spread that we saw, and the fact that it spread to new rooms, suggest that Ebola aerosols were being generated and were present in the building," he said. "If you look at pictures of lungs from a monkey with Ebola Zaire, you see that the lungs are fogged with Ebola. Have you seen those pictures?"

"Yes. Nancy Jaax showed them to me."

"Then you know. You can see Ebola particles clearly in the air spaces of the lung."

"Did you ever try to see if you could put Ebola Reston into the air and spread it among monkeys that way?" I asked.

"No," he replied firmly. "I just didn't think that was a good idea. If anybody had found out that the Army was doing experiments to see if the Ebola virus had adapted to spreading in the respiratory tract, we would have been accused of doing offensive biological warfare—trying to create a doomsday germ. So we elected not to follow it up."

"That means you don't really know if Ebola spreads in the air."

"That's right. We don't know. You have to wonder if Ebola virus can do that or not. If it can, that's about the worst thing you can imagine."

So the three sisters—Marburg, Ebola Sudan, and Ebola Zaire—have been joined by a fourth sister, Reston. A group of researchers at the Special Pathogens Branch of the C.D.C. —principally Anthony Sanchez and Heinz Feldmann—have picked apart the genes of all the filoviruses. They've discovered that Zaire and Reston are so much alike that it's hard to say how they are different. When I met Anthony Sanchez and asked him about it, he said to me, "I call them kissing cousins. But I can't put my finger on why Reston apparently doesn't make us sick. Personally, I wouldn't feel comfortable handling it without a suit and maximum containment procedures."

Each filovirus strain contains seven proteins, four of which are completely unknown. Something slightly different about one of the Reston proteins is probably the reason the virus didn't go off in Washington like a bonfire. The Army and the C.D.C. have never downgraded the safety status of Reston virus. It remains classified as a Level 4 hot agent, and if you want to shake hands with it, you had better be wearing a space suit. Safety experts feel that there is not enough evidence, yet, to show that the Reston strain is not an extremely dangerous virus. It may be, in fact, the most dangerous of all the filovirus sisters, because of its seeming ability to travel rather easily through the air, perhaps more easily than the others. A tiny change in its genetic code, and it might turn into a cough and take out the human race.

Why is the Reston virus so much like Ebola Zaire, when Reston supposedly comes from Asia? If the strains come from different continents, they should be quite different from each other. One possibility is that the Reston strain originated in Africa and flew to the Philippines on an airplane not long ago. In other words, Ebola has already entered the net and has been traveling lately. The experts do not doubt that a virus can hop around the world in a matter of days. Perhaps Ebola came out of Africa and landed in Asia a few years back. Perhaps—this is only a guess—Ebola traveled to Asia inside wild African animals. There have been rumors that wealthy people in the Philippines who own private estates in the rain forest have been importing African animals illegally, releasing them into the Philippine jungle, and hunting them. If Ebola lives in African game animals—in leopards or lions or in Cape buffalo—it might have traveled to the Philippines that way. This is only a guess. Like all the other thread viruses, Ebola Reston hides in a secret place. It seems quite likely, however, that the entire Reston outbreak started with a single monkey in the Philippines. One sick monkey. That monkey was the unknown index case. One monkey started the whole thing. That monkey perhaps picked up four or five particles of Ebola that came from . . . anyone's guess.

PART FOUR

KITUM CAVE

Highway

The road to Mount Elgon heads northwest from Nairobi into the Kenya highlands, climbing through green hills that bump against African skies. It goes through small farms and patches of cedar forest, and then it breaks over a crest of land and seems to leap out into space, into a bowl of yellow haze, which is the Rift Valley. The road descends into the Rift, cutting across wrinkled knees of bluffs, until it hits bottom and unravels on a savanna dotted with acacia trees. It skirts the lakes at the bottom of the Rift and passes through groves of fever trees, yellow-green and glowing in the sun. It is detained in cities that dwell by the lakes, and then it turns westward, toward a line of blue hills, which is the western side of the Rift, and it climbs into the hills, a straight, narrow, paved two-lane highway, crowded with smoky overlander trucks gasping up the grade, bound for Uganda and Zaire.

The road to Mount Elgon is a segment of the AIDS highway, the Kinshasa Highway, the road that cuts Africa in half, along which the AIDS virus traveled during its breakout from somewhere in the African rain forest to every place on earth. The road was once a dirt track that wandered through the heart of Africa, almost impossible to traverse along its complete length. Long sections of it were paved in the nineteen-seventies, and the trucks began rolling through, and soon afterward the AIDS virus appeared in towns along the highway. Exactly where the virus came from is one of the great mysteries.

The road to Mount Elgon was familiar to me; I had traveled over it as a boy. My parents and my brothers and I had lived for a short while with a Luo family on their farm in the hills overlooking Lake Victoria—a traditional farm, with mud huts and a *boma* for keeping cattle. I had not been back to Africa since I was twelve years old, but when you have encountered Africa in childhood, it becomes a section of your mind. I had felt warm river sand on my bare feet and had smelled crocodiles. I knew the crispy sensation of tsetse flies crawling in my hair. I could still hear the sound of voices speaking English in the soft accents of the Luo language, urging me to feel free, feel free, eat more fat from the ram's tail. I knew what it felt like to wake up in gray light before dawn not knowing where I was, seeing a mud wall with a hole in it, and gradually realizing that the hole was a window in a hut and that I was being watched through the window by a crowd of children. When I saw Africa again, Africa came back whole, alive, shining with remembered enigma. What came back first was the smell of Africa, the smoky smell of cooking fires, which produce a haze of burning acacia and blue-gum wood that covers the towns and clings to the bodies of people. What came back to me next, with a slap of recognition, was the sight of crowds of people walking along the roads as if they had been walking since the beginning of time, heading nowhere and everywhere by foot. In the highlands of Kenya, their bare and sandaled feet pound the shoulders of the highway into braids of red clay. The women sing Christian hymns as they walk, and some carry guitars, or they carry sacks of charcoal or salt balanced on their heads.

• • •

The Land Rover smashed through a pall of diesel smoke and
bounced as it hit a pothole. Robin MacDonald, my guide,
gripped the steering wheel. "Oh, this road is good, man," he
said approvingly. "Last time I was here, it was so bad you
would be *crying* by now. I haven't been up to Mount Elgon
in years—not since I was a kid, really. My old man had a
friend who had a shamba up there"—*shamba,* a farm—
"and we used to visit him all the time. Oh, it was lovely,
man. That farm is gone now. Eh, it's *kwisha.*" *Kwisha:*
finished. He dodged around a herd of goats, using his horn
liberally. "Get out of the way, man!" he shouted at a goat.
"Look, he's not even moving." The Land Rover roared and
accelerated.

The road passed through small cornfields. In the middle
of each plot stood a hut made of mud or cement. People
stooped among the cornstalks, hand-tilling their fields with
mattocks. Every inch of land lay under cultivation, right up
to the doors of the houses. We passed a man standing by the
road, holding a suitcase tied up with string. He waved to us.
We passed another man wearing an English raincoat and a
fedora hat and carrying a stick, walking slowly: a gray figure
in bright sun. Some people waved as we passed, and others
turned around in their tracks and stared at us. We stopped to
wait for a herd of cattle crossing the road, driven by Kikuyu
boys holding switches.

"Ay," Robin said dreamily. "When I was a kid, this
country was different, eh? To get anywhere in this country
was a three-day trip. We shot a bloody Thomson's gazelle
and lived off the thing the whole time. In the old days,
twenty years ago, this land was all forest and grassland. Now
it's corn. Everywhere corn. And the forests are gone, man."

Robin MacDonald is a professional hunter and safari
guide. He is one of about two dozen professional hunters
who are left in East Africa. They take clients into the bush to

hunt big game. He has a broad, ruddy face, thin lips, piercing eyes behind eyeglasses, and broad cheekbones. He has black, curly hair that hangs in pieces around his forehead, looking like he'd chopped it off with a knife. For walking in the bush, he wears a baseball cap, a black T-shirt, shorts, a curved African knife at his belt, and scorched, melted green sneakers—dried too many times over campfires. He is the son of a famous professional hunter named Iain MacDonald, who was killed at the controls of a light plane that crashed on the African plains in 1967, when Robin was thirteen. By that time, Robin had learned what he needed to know. He had hunted leopard and lion with his father, and he had already shot his first charging Cape buffalo while his father stood beside him to make the back-up shot in case he missed. Robin tracked elephant with his father for days through the dry thornbush of the Yatta Plateau, carrying nothing but a canteen of water and one apple—"That client, he was a guy from Texas, that guy," Robin explained. "He said he could walk it no problem, said he was an experienced hunter. He sat down one day and said, 'To hell with this, I can't go on. Make me a camp.' So we made him a camp, and we went on, my old man and I, and we stalked the elephant for two days. My old man only took water when he was tracking an elephant. Said to me, 'Stuff an apple in that pack, and we'll be off.' And then we walked across the Yatta Plateau for two days. When we found the elephant, we led the client to it, and he shot it."

"How old were you then?"

"Seven, man."

He does not hunt elephants anymore—he approves of the current worldwide ban on ivory—but he does hunt Cape buffalo, which is not an endangered species.

There had been reports of tribal violence around Mount Elgon. The Elgon Masai had been raiding the Bukusu, an ethnic group whose people live on the southern side of the mountain, and had been burning their huts, shooting them with automatic weapons, and driving them off their land. I was concerned about the situation and had telephoned Robin from the United States to ask him his opinion.

"Where do you want to go? Mount Elgon?" he had said. His voice sounded hissy and remote on the line.

"I'm bringing a couple of space suits with me," I said.

"Whatever, my man."

"Is it safe to travel around Mount Elgon?"

"No hassle. Not unless there's a bloody uproar."

He lighted a cheap African cigarette and glanced at me. "So what are your plans for the cave? Are you going to collect any *specimens?* Any boxes of bat shit or the like?"

"No, I just want to look around."

"I used to go up to that cave when I was a kid," he said. "So there's a disease up there, eh? Makes AIDS look like a sniffle, eh? You turn into soup, eh? You explode, eh? *Pfft!*— coming out of every hole, is that the story? And how long does it take?"

"About seven days."

"Oof! Man. How do you get it?"

"By touching infected blood. It may also be airborne. It is also sexually transmitted."

"Like AIDS, you mean?"

"Yes. The testicles swell up and turn black-and-blue."

"What! Your guliwackers blow up? Lovely! So you get balls like a blue monkey! Christ! That virus is a bloody *shit,* that one."

"You have given a good description of the agent," I said.

Robin breathed his cigarette. He removed the baseball cap from his head and smoothed his hair and replaced the cap. "Right, then. You'll go inside the cave and have a look at bat shit. And then—and then—after you *explode* in one of my tents, what shall I do with you?"

"Don't touch me. You could get very sick if you touch me. Just roll up the tent with me inside it, and take the thing to a hospital."

He crinkled with laughter. "Right. We'll call in the Flying Doctors. They'll pick up anything. And which hospital shall we have you delivered to, eh?"

"Nairobi Hospital. Leave me by the Casualty entrance."

"Right, my man. That's what we'll do."

The Cherangani Hills appeared in the distance, a line of mountains on the edge of the Rift, humpy and green, crushed under an indwelling sweep of rain clouds. The clouds darkened and gathered together as we approached Mount Elgon, and splats of rain began to hit the windshield. The air turned cold and raw. Robin turned on his headlights.

"Did you find some bleach?" I asked him.

"I've got a gallon in back."

"Plain laundry bleach?"

"Right. We call it Jik here in Kenya. Bloody Jik."

"Is it like Clorox?"

"Right. Jik. Drink it, and it will bloody kill you."

"I hope it kills Marburg."

The country grew more settled, and we passed through towns. Everywhere we saw overlander trucks parked in front of shacks made of planks and metal. They were small restaurants. Some of them were full-service establishments, offering grilled goat, Tusker beer, a bed, and a woman. Medical doctors who work in East Africa believe that 90 percent of the prostitutes working along the main roads carry the AIDS virus. No one knows for sure, but local doctors think that as many as 30 percent of all men and women of childbearing age who live in the vicinity of Mount Elgon are infected with HIV. Most of them will die of AIDS. Many of their newborn children will also contract AIDS and die of the virus in childhood.

The emergence of HIV was subtle: it incubates for years in a human host before it kills the host. If the virus had been

noticed earlier, it might have been named Kinshasa Highway, in honor of the fact that it passed along the Kinshasa
Highway during its emergence from the African forest.
When I rode along the Kinshasa Highway as a boy, it was a
dusty, unpaved thread that wandered through the Rift Valley
toward Lake Victoria, carrying not much traffic. It was a
gravel road engraved with washboard bumps and broken by
occasional pitlike ruts that could crack the frame of a Land
Rover. As you drove along it, you would see in the distance a
plume of dust growing larger, coming toward you: an automobile. You would move to the shoulder and slow down, and
as the car approached, you would place both hands upon the
windshield to keep it from shattering if a pebble thrown up
by the passing car hit the glass. The car would thunder past,
leaving you blinded in yellow fog. Now the road was paved
and had a stripe painted down the center, and it carried a
continual flow of vehicles. The overlanders were mixed up
with pickup trucks and vans jammed with people, and the
road reeked of diesel smoke. The paving of the Kinshasa
Highway affected every person on earth, and turned out to be
one of the most important events of the twentieth century. It
has already cost at least ten million lives, with the likelihood
that the ultimate number of human casualties will vastly
exceed the deaths in the Second World War. In effect, I had
witnessed a crucial event in the emergence of AIDS, the transformation of a thread of dirt into a ribbon of tar.

Camp

Robin's wife, Carrie MacDonald, is his business partner, and she often accompanies him on safaris with clients. The MacDonalds also bring along their two small sons, if the client will allow that. Carrie is in her twenties, with blond hair and brown eyes and a crisp English accent. Her parents brought her to Africa from England when she was a girl.

We traveled in two Land Rovers, Carrie driving one and Robin driving the other. "We always take two vehicles in this country, in case one breaks down," Carrie explained. "It happens literally all the time." Carrie and Robin's two boys rode with Carrie. We were also accompanied by three men who were members of the MacDonalds' safari staff. Their names are Katana Chege, Herman Andembe, and Morris Mulatya. They are professional safari men, and they do most of the work around the campsite. They spoke very little English and had résumés as long as one's arm. In addition to those people, two friends of mine had joined the expedition. One was a childhood friend named Frederic Grant, and the other was a woman named Jamy Buchanan; both are Americans. I had prepared a written list of instructions for my friends in case I broke with Marburg, and I had sealed the document in an envelope and hidden it in my backpack. It ran for three pages, typewritten, single spaced, describing the signs and symptoms of a filovirus infection in a human being, as well as possible experimental treatments that might arrest the terminal meltdown. I had not told my friends about this envelope, but I planned to give it to them if I came down with a headache. This was a sign of nervousness, to say the least.

Robin turned into the opposite lane in order to pass a truck, and suddenly we were headed straight for an oncoming car. Its headlights flashed and its horn wailed.

Fred Grant grabbed the seat and shouted, "Why is this guy coming at us?"

"Yeah, well, we're going to die, so don't worry about it," Robin remarked. He dodged in front of the truck just in time. He blurted out a song:

> Livin' and a-lovin'
> And a-lovin' and a-livin'—Yah!

We stopped and bought roasted ears of corn from a woman standing by the side of the road with a charcoal brazier. The corn was hot, dry, scorched, and delicious, and it cost five cents. It was called a mealy.

Robin chewed his mealy as he drove. Suddenly he grabbed his jaw and swore violently. "My tooth! Bloody hell! A filling came out! This arsehole of a bloody dentist!" He rolled down his window and spat corn and bits of metal filling into the wind. "Well, carry on. Three fillings, and they've *all* come out now. Carrie sent me to the man. Said he was a good dentist—hah!"

He floored his Land Rover until it hovered behind Carrie's Land Rover. The two vehicles were roaring down the highway as if they were attached to each other. He leaned out the window and hurled his gnawed mealy at his wife's Land Rover. It bounced off her rear window. She didn't seem to notice. We passed a sign that said: REDUCE ROAD CARNAGE— DRIVE SAFELY.

Toward sunset, we stopped in the town of Kitale, at the base of Mount Elgon, to buy Tusker beers and charcoal.

Kitale is a market town. The main market is situated along the highway leading into town, near an old train station built by the English. The highway is lined with towering blue-gum trees. Under the trees, on pounded dirt and among mud puddles from fresh rains, people set up stands for selling umbrellas and plastic wrist watches. Robin turned his Land Rover into the market and drove slowly through the crowds. A man shouted in Swahili, "You are driving the wrong way!"

"Where are the signs?" Robin shouted back.

"We don't need signs here!"

We parked and walked through the town, and instantly we were surrounded by pimps. One guy wore a white ski parka and said, "Do you want to go Kigawera? Yes? I will take you there. Come with me. Right now. Beautiful girls. I will take you there." That might be the neighborhood where Charles Monet's girlfriends had lived, who knows. It was rush hour, and crowds flowed on foot under the gum trees, past an endless line of small shops. Mount Elgon brooded over the town and the trees, rising to an undefined height, its profile buried in an anvil thunderhead, bathed in golden light. An edge of the mountain razored diagonally upward into the cloud. A silent flash traveled around the mountain, followed by another flash—chain lightning, but no thunder reached the town. The air was cold, heavy, wet, and filled with the sound of crickets.

In our explorations on mud roads around Mount Elgon, we saw signs of the recent trouble: burned, empty huts that had once belonged to Bukusu farmers. Someone had warned me that we would hear gunfire at night, but we didn't. Sickly banana trees leaned around the abandoned huts. The huts stood in fallow fields, studded with African weeds and shoots of young saplings. We made a camp in the same meadow where Charles Monet had camped. The cook, Morris Mulatya, dumped a sack of charcoal on the ground and built a fire, and put a metal teapot on it to heat water for tea. Robin MacDonald sat down on a folding chair and removed his sneakers. He rubbed his feet with his hands and then drew his knife from its sheath and began paring bits of skin from his toes. Not far away, at the edge of the forest that ringed our campsite, a Cape buffalo eyed us. Robin eyed the buffalo. "That's a male," he muttered. "Those are bastards. You've got to watch them. They'll *lift* you. The Cape buffalo have killed more human beings in Africa than any other animal. Except hippo. Those swine have killed more."

I knelt in the grass and organized a row of boxes that contained space suits, decontamination gear, and lights. Smoke from the campfire curled in the air, which was filled with the *clink-clank* noise of safari tents being erected by the MacDonalds' staff. Carrie MacDonald worked around the campsite, getting things organized, speaking Swahili to the men. A nearby stream tumbled out of a glade. Robin looked up, listening to birds. "Hear that? Those are turacos. And there's a wood hoopoe. And there's a gray mousebird; do you see that long tail?"

He wandered down to the stream. I followed him. "I wonder if there's any trout in here," he said, staring into the water. "This could be good for fly fishing."

I put my hand into the water. It was ice-cold and bubbly but gray in color, clouded with volcanic dust, not the kind of water that would sustain trout.

"Talk about fly fishing. Did you ever hear of fly fishing for crocodiles?" Robin said.

"No."

"You put a piece of meat on a chain. A piece of meat *this* big. And the flies are all *over* the place! Now there's some fly fishing! They stink, those crocodiles. You'll be standing in shallow water, and they'll swim up on you. And the water is muddy. And you can't see them. And unless you can smell them, you don't know they're there. And then—*pfft!* They drag you down. End of story. You're history, my man. Talk about Nature. The whole thing, if you think about it, is full of killers, from the river to the sea.''

A young man in a beret and military fatigues knelt on one knee in the grass, holding a Russian assault rifle, watching us with mild interest. His name was Polycarp Okuku, and he was an *askari,* an armed guard.

"Iko simba hapa?" Robin called to him. Any lions around here?

"Hakuna simba." No lions left.

Poachers from Uganda had been coming over Mount Elgon and shooting anything that moved, including people, and now the Kenyan government required that visitors to Mount Elgon be accompanied by armed guards. The Swahili word *askari* used to mean "spear bearer." Now it means a man who carries an assault rifle and who walks in your shadow.

Kitum Cave opens in a forested valley at an altitude of eight thousand feet on the eastern slope of the mountain. *"Whoof!"* MacDonald said as we grunted up the trail. "You can smell the Cape buffalo around here, eh? *Mingi* buffalo." *Mingi:* many. Lots of buffalo. Buffalo trails crossed the human trail on diagonals. The trails were wider, deeper, straighter, more businesslike than the human trail, and they reeked of buffalo urine.

I was wearing a backpack. I picked my way across muddy spots in the trail.

Polycarp Okuku yanked a lever on the barrel of his assault rifle, *clack, ta chock.* This action cocked the weapon and slotted a round into the firing chamber. "Especially in the rainy season, the Cape buffalo like to travel in herds," he explained.

The sound of a machine gun being cocked brought Robin to attention. "Bloody hell," he muttered. "That toy he's carrying isn't safe."

"Look," Okuku said, pointing to a clump of boulders. "Hyrax." We watched a brown animal about the size of a

woodchuck run silkily down the rocks. A possible host of
Marburg virus.

The valley was cloaked in African olive trees, African
cedars, broad-leaved croton trees, *Hagenia abyssinica* trees
drenched in moss, and whiplike young gray Elgon teaks.
Here and there grew an occasional podocarpus tree, with a
straight, silvery shaft that thrust upward to an incredible
height and vanished in the shifting green of biological space.
This was not lowland rain forest, where the crowns of trees
merge into a closed canopy, but an African montane rain
forest, a particular kind of forest with a broken canopy, penetrated by holes and clearings. Sunlight fell in shafts to the
forest floor, washing over glades where nettles and papyrus
sparkled with wild violets. Each tree stood in a space of its
own, and the branches zigzagged against the clouds and sky
like arms reaching out for heaven. From where we stood, we
could see farms on the lower slopes of the mountain. As the
eye moved from the lowlands to the uplands, the farms gave
way to patches of shrubby trees, to fingers and clumps of
larger trees, and then to an unbroken blanket of primeval
East African rain forest, one of the rarest and most endangered tropical forests on the planet.

The color of the forest was a silvery gray-green from the
olive trees, yet here and there, a dark-green podocarpus tree
burst through the canopy. A podo tree's shaft is lightly fluted
and goes straight up, without branches, sometimes spiraling
as it goes, and there may be a slight swing or curve in the
shaft, which gives the tree a look of tension and muscularity,
like a bent bow. High up, the podo tree flares into a vase-
shaped crown, like an elm tree, and the downhanging limbs
are draped with bundles of evergreen leafy needles and are
spangled with ball-shaped fruit. The podos were hard to see
in the thickets near Kitum Cave, because they did not grow
large in that valley, but I noticed a young podo that was
seven feet thick and close to a hundred feet tall. I guessed it
had begun to grow in the time of Beethoven.

"What's missing here is the game," Robin said. He stopped and adjusted his baseball cap, surveying the forest. "The elephants have been all shot to shit. If they hadn't been shot, man, you'd see them all over this mountain. *Mingi* elephant. This whole *place* would be elephant."

The valley was quiet, except for the remote *huh, huh* of colobus monkeys that retreated from us as we climbed. The mountain seemed like an empty cathedral. I tried to imagine what it must have been like when herds of elephants could have been seen moving through a forest of podo trees as large as sequoias: only ten years ago, before the trouble, Mount Elgon had been one of the earth's crown jewels.

The mouth of Kitum Cave was mostly invisible from the approach trail, blocked by boulders cloaked in moss. A choir of African cedar trees grew in a row over the mouth of the cave, and a small stream trickled down among the cedars and rained over the boulders, filling the valley with a sound of falling water. As we got closer, the sound of the waterfall grew louder, and the air began to smell of something alive. It smelled of bat.

Giant stinging nettles grew in clumps among the boulders, and they brushed against our bare skin and caught our legs on fire. It occurred to me that nettles are, in fact, injection needles. Stinging cells in the nettle inject a poison into the skin. They break the skin. Maybe the virus lives in nettles. Moths and tiny flying insects drifted out of the cave mouth, carried in a steady, cool flow of air. The insects floated like snow blown sideways. The snow was alive. It was a snow of hosts. Any of them might be carrying the virus, or none of them.

We stopped on an elephant trail that led into the cave, beside a wall of rock that was covered with diagonal hatch marks made by the elephants tusking the rocks for salt. The forests of Mount Elgon were home to two thousand elephants, until the men with machine guns came over from Uganda. Now the Mount Elgon herd has withered to one extended family of about seventy elephants. The poachers set up a machine-gun nest at the mouth of Kitum Cave, and after that the surviving elephants had learned their lesson. The herd stays out of sight as much as possible, concealed in valleys higher on the mountain, and the smart old females, the grandmothers, who are the bosses of the herd and who direct its movements, lead the others to Kitum Cave about once every two weeks, when the elephants' hunger for salt overcomes their fear of being shot.

Elephants had not been the only visitors to Kitum Cave. Cape buffalo had gouged footprints in the trail leading into the cave. I noticed fresh, green splats of buffalo dung, and waterbuck hoofprints. The trail itself seemed to consist of a bed of dried animal dung. Other than the elephant herd, many different kinds of animals had been going inside Kitum Cave—bushbuck, red duikers, perhaps monkeys, perhaps baboons, and certainly genet cats, which are wild cat-like animals somewhat larger than a house cat. Rats, shrews, and voles go inside the cave, too, either looking for salt or foraging for food, and these small mammals make trails through the cave. Leopards go inside the cave at night, looking for prey. Kitum Cave is Mount Elgon's equivalent of the Times Square subway station. It is an underground traffic zone, a biological mixing point where different species of animals and insects cross one another's paths in an enclosed air space. A nice place for a virus to jump species.

I unzipped my backpack and withdrew my gear and laid it on the rocks. I had assembled the components of a Level 4 field biological space suit. It was not a pressurized suit—not an orange Racal suit. It was a neutral-pressure whole-body suit with a hood and a full-face respirator. The suit itself was made of Tyvek, a slick, white fabric that is resistant to moisture and dust. I laid out a pair of green rubber gauntlet gloves, yellow rubber boots, a black mask with twin purple filters. The mask was a silicone rubber North respirator mask with a Lexan faceplate, for good visibility, and the purple filters were the kind that stop a virus. The mask had an insectile appearance, and the rubber was black and wet looking, sinister. I placed a roll of sticky tape on the rocks. A plastic shower cap—ten cents apiece at Woolworth's. Flashlight, head lamp. I stepped into the suit, feet first, pulled it up to my armpits, and fed my arms into the sleeves. I stretched the shower cap over my head and then pulled the hood of the suit down over the shower cap. I zipped up the front zipper of the suit, from crotch to chin.

Generally you need a support team to help you put on a field biological suit, and my traveling companion Fred Grant was acting in this capacity. "Could you hand me the sticky tape?" I said to him.

I taped the front zipper of the suit, taped the wrists of my gloves to the suit, taped the cuffs of the boots to the suit.

Polycarp Okuku sat on a rock with his gun across his knees, gazing at me with a carefully neutral expression on his face. It was evident that he did not want anyone to think he was surprised that someone would put on a space suit to go inside Kitum Cave. Later he turned and spoke at length in Swahili with Robin MacDonald.

Robin turned to me. "He wants to know how many people have died in the cave."

"Two," I said. "Not in the cave—they died afterward. One was a man, and the other was a boy."

Okuku nodded.

"There's very little danger," I said. "I'm just being careful."

Robin scuffed his sneaker in the dirt. He turned to the *askari* and said, "You explode, man. You get it, and that's it —*pfft!*—end of story. You can kiss your arse good-bye."

"I have heard about this virus," Okuku said. "There was something the Americans did at this place."

"Were you working here then?" I asked. When Gene Johnson and his team came.

"I was not here then," Okuku said. "We heard about it."

I fitted the mask over my face. I could hear my breath sucking in through the filters and hissing out through the mask's exhaust ports. I tightened some straps around my head.

"How does it feel?" Fred asked.

"Okay," I said. My voice sounded muffled and distant to my ears. I inhaled. Air flowed over the faceplate and cleared it of fog. They watched me fit an electric miner's lamp over my head.

"How long are you going to be in there?" Fred asked.

"You can expect me back in about an hour."

"An hour?"

"Well—give me an hour."

"Very well. And then?" he asked.

"And then? Dial 9-1-1."

The entrance is huge, and the cave widens out from there. I crossed a muddy area covered with animal tracks and continued along a broad platform covered with spongy dried dung. With the mask over my face, I could not smell bats or dung. The waterfall at the cave's mouth made splashing echoes. I turned and looked back, and saw that clouds were darkening the sky, announcing the arrival of the afternoon rains. I turned on my lights and walked forward.

Kitum Cave opens into a wide area of fallen rock. In 1982, a couple of years after Charles Monet visited the cave, the roof fell in. The collapse shattered and crushed a pillar that had once seemed to support the roof of the cave, leaving a pile of rubble more than a hundred yards across, and a new roof was formed over the rubble. I carried a map inside a plastic waterproof bag. The bag was to protect the map, to keep it from picking up any virus. I could wash the bag in bleach without ruining the map. The map had been drawn by an Englishman named Ian Redmond, an expert on elephants who once lived inside Kitum Cave for three months, camping beside a rock near the entrance while he observed the elephants coming and going at night. He wore no biohazard gear and remained healthy. (Later, when I told Peter Jahrling of USAMRIID about Redmond's camp-out inside Kitum Cave, he said to me, in all seriousness, ''Is there any way you could get me a little bit of his blood, so we can run some tests on it?'')

It was Ian Redmond who conceived the interesting idea that Kitum Cave was carved by elephants. Mother elephants teach their young how to pry the rocks for salt—rock carving is a learned behavior in elephants, not instinctive, taught to children by their parents; this knowledge has been passed down through generations of elephants for perhaps hundreds of thousands of years, for perhaps longer than modern humans have existed on the earth. If the elephants have been tusking out the rock of Kitum Cave at a rate of a few pounds a night, the cave could easily have been carved by elephants over a few hundred thousand years. Ian Redmond figured this out. He calls it speleogenesis by elephants—the creation of a cave by elephants.

The light began to fade, and the mouth of the cave, behind me, became a crescent of sunlight against the high, fallen ceiling. Now the mouth looked like a half-moon. I came to a zone of bat roosts. These were fruit bats. My lights disturbed them, and they dropped off the ceiling and flitted past my head, giving off sounds that resembled Munchkin laughter. The rocks below the bats were slubbered with wet, greasy guano, a spinach-green paste speckled with gray blobs, which reminded me of oysters Rockefeller. Momentarily and unaccountably, I wondered what the bat guano would taste like. I thrust away this thought. It was the mind's mischief. You should avoid eating shit when you are in Level 4.

Beyond the bat roosts, the cave became drier and dustier. A dry, dusty cave is very unusual. Most caves are wet, since most caves are carved by water. There was no sign of running water in this cave, no streambed, no stalactites. It was an enormous, bone-dry hole in the side of Mount Elgon. Viruses like dry air and dust and darkness, and most of them don't survive long when exposed to moisture and sunlight. Thus a dry cave is a good place for a virus to be preserved, for it to lie inactive in dung or in drying urine, or even, perhaps, for it to drift in cool, lightless, nearly motionless air.

Marburg-virus particles are tough. One would imagine they can survive for a fair amount of time inside a dark cave. Marburg can sit unchanged for at least five days in water. This was shown by Tom Geisbert. One time, just to see what would happen, he put some Marburg particles into flasks of room-temperature water and left the flasks sitting on a countertop for five days (the counter was in Level 4). Then he took the water and dropped it into flasks that contained living monkey cells. The monkey cells filled up with crystalloids, exploded, and died of Marburg. Tom had discovered that five-day-old Marburg-virus particles are just as lethal and infective as fresh particles. Most viruses do not last long outside a host. The AIDS virus survives for only a few minutes when exposed to air. No one has ever tried to see how long Marburg or Ebola can survive while stuck to a dry surface. Chances are the thread viruses can survive for some time—if the surface is free of sunlight, which would break apart the virus's genetic material.

I came to the top of the mound, reached out with my gloved hand, and touched the ceiling. It was studded with brown oblong shapes—petrified tree logs—and whitish fragments—pieces of petrified bone. The rock is solidified ash, the relic of an eruption of Mount Elgon. It is embedded with stone logs, the remains of a tropical rain forest that was swept up in the eruption and buried in ash and mud. The logs are dark brown and shiny, and they reflected opalescent colors in the beam of my head lamp. Some of the logs had fallen out of the roof, leaving holes, and the holes were lined with white crystals. The crystals are made of mineral salts, and they looked evilly sharp. Had Peter Cardinal reached up and touched these crystals? I found bats roosting in the holes

among the crystals—insect-eating bats, smaller than the fruit
bats that clustered near the cave's mouth. As I played my
head lamp over the holes, bats exploded out of them and
whirled around my head and were gone. Then I saw something wonderful. It was the tooth of a crocodile, caught in the
rock. The ash flow had buried a river that had contained
crocodiles. The crocodiles had been trapped and burned to
death in an eruption of Mount Elgon. Full of killers, from
the river to the sea.

I shuffled across razorlike slices of rock that had fallen
from the roof, and came to a fresh elephant dropping. It was
the size of a small keg of beer. I stepped over it. I came to a
crevice and shone my lights down into it. I didn't see any
mummified baby elephants down there. I came to a wall. It
was scored with hatch marks—elephant tuskings. The elephants had left scrapes in the rock all over the place. I kept
going down and came to a broken pillar. Next to it, a side
tunnel continued downward. I wormed into the passage, on
my knees. It circled around and came out in the main room. I
was boiling hot inside the suit. Drops of moisture had collected on the inside of my faceplate and pooled in the mask
under my chin. My footsteps kicked up dust, and it rose in
puffs around my boots. It felt strange to be soaking wet and
yet wading through dust. As I was climbing out of the passage, my head slammed against a rock. If I hadn't been
wearing protective gear, the rock would have cut my scalp. It
seemed easy to get a head wound in the cave. Perhaps that
was the route of infection: the virus clings to the rocks and
gets into the bloodstream through a cut.

I proceeded deeper until I came to a final wall in the
throat of the cave. There, at knee level, in total darkness, I
found spiders living in webs. They had left their egg cases
scattered about, hanging from the rock. The spiders were
carrying on their life cycle at the back of Kitum Cave. That
meant they were finding something to eat in the darkness,

something that was flying into their webs. I had seen moths
and winged insects pouring from the mouth of the cave, and
it occurred to me that some of them must be flying all the
way to the back. The spiders could be the host. They could
catch the virus from an insect in their diet. Perhaps Marburg
cycles in the blood of spiders. Perhaps Monet and Cardinal
were bitten by spiders. You feel a cobweb clinging to your
face and then comes a mild sting, and after that you don't
feel anything. You can't see it, you can't smell it, you can't
feel it. You don't know it's there until you start to bleed.

So much was happening that I didn't understand. Kitum
Cave plays a role in the life of the forest, but what the role is
no one can say. I found a crevice that seemed to be full of
clear, deep water. It couldn't be water, I thought, the crevice
must be dry. I picked up a stone and threw it. Halfway down
in its flight, the stone made a splash. It had hit water. The
stone spun lazily downward into the crevice and out of sight,
and ripples spread across the pool and died away, throwing
reflections of my head lamp onto the wall of the cave.

I climbed over fallen plates of stone back to the top of the
rubble pile, playing my lights around. The room was more
than a hundred yards across, larger in all directions than a
football field. My lights failed to penetrate to the edges of
the room, and the edges descended downward into darkness
on all sides. The mound of rubble in the center made the
cave resemble the curving roof of a mouth. As you look into
someone's mouth, you see the tongue in front, lying under
the roof of the mouth, and you see the tongue curving backward and down into the throat: that is what Kitum Cave
looked like. Say *"Ahh,"* Kitum Cave. Do you have a virus?
No instruments, no senses can tell you if you are in the
presence of the predator. I turned off my lights and stood in
total darkness, feeling a bath of sweat trickle down my chest,
hearing the thump of my heart and the swish of blood in my
head.

• • •

The afternoon rains had come. Fred Grant was standing inside the mouth of the cave to keep himself dry. The *askari* sat on a rock nearby, bouncing the machine gun on his knees, looking bored.

"Welcome back," Grant said. "Was it good for you?"

"We'll find out in seven days," I said.

He scrutinized me. "There appear to be splatters on your face shield."

"Splatters of *what?*"

"Looks like water."

"It's just sweat inside my mask. If you'll bear with me a moment, I'm going to get this suit off." I took a plastic laundry tub—part of the gear we had carried up to the cave —and left it under the waterfall for a moment. When the tub was partly full, I carried it over to the elephants' pathway, at the entrance, put it on the ground, and poured in most of a gallon of "bloody Jik"—laundry bleach.

I stepped into the tub. My boots disappeared in a swirl of dirt coming off them, and the Jik turned brown. I put my gloved hands into the brown Jik, scooped up some of the liquid, and poured it over my head and face mask. Using a toilet brush, I scrubbed my boots and legs to remove obvious patches of dirt. I dropped my bagged map into the Jik. I dropped my flashlight and head lamp into the Jik. I took off my face mask and dunked it, along with the purple filters. Then my eyeglasses went into the Jik.

I peeled off my green gauntlet gloves. They went into the Jik. I stepped out of my Tyvek suit, peeling the sticky tape as I went. The whole suit, together with the yellow boots, went into the Jik. It was a stew of biohazard gear.

Underneath my suit, I wore a set of clothes and a pair of sneakers. I stripped to the skin and put the clothing into a plastic garbage bag—a so-called hot bag—along with a splash or two of Jik, and then put that bag into another bag. I washed the outsides of both bags with bleach. From my backpack, I removed a clean set of clothing and put it on. I put the biohazard gear into double bags, adding Jik.

Robin MacDonald appeared noiselessly in his sneakers at the top of the rocks at the mouth of the cave. "Sir Bat Shit!" he called. "How did it go?"

We walked down the trail, lugging the hot bags, and returned to camp. The rain intensified. We settled down on chairs in the mess tent with a bottle of scotch whisky, while the rain splattered down and hissed through the leaves. It was three o'clock in the afternoon. The clouds thickened to the point where the sky grew black, and we lighted oil lamps inside the mess tent. Peals of thunder rolled around the mountain, and the rain turned into a downpour.

Robin settled into a folding chair. "Ah, man, this rain never stops on Elgon. This happens all year round."

There was a stroboscopic flash and a bang, and a lightning bolt whacked an olive tree. The flash outlined his face, his glasses. We chased the scotch with Tusker beers and played a round of poker. Robin declined to join the game. I had the impression he didn't know how to play poker.

"Have some whisky, Robin," Fred Grant said to him.

"None of that for me," he said. "My stomach doesn't like it. Beer is just right. It gives you protein, and you sleep well."

The rain tapered off, and the clouds momentarily lightened. Olive trees arched overhead in squiggles, their feet sunk in shadows. Water drops fell through the halls of trees. Mousebirds gave off flutelike cries, and then the cries stopped, and Mount Elgon became silent. The forest shifted gently, rocking back and forth. Rain began to fall again.

"How are you feeling, Sir Bat Shit?" Robin said. "Are you getting any mental symptoms? That's when you first start talking to yourself in the toilet. It'll be starting any day now."

The mental symptoms were starting already. I remembered slamming my head into the roof of the cave. That had raised a bump on my scalp. There would be microscopic tears in the skin around that bump. I had begun to understand the feeling of having been exposed to a filovirus: I'll be okay. No problem. The odds are very good that I wasn't exposed to anything.

The emergence of AIDS, Ebola, and any number of other rainforest agents appears to be a natural consequence of the ruin of the tropical biosphere. The emerging viruses are surfacing from ecologically damaged parts of the earth. Many of them come from the tattered edges of tropical rain forest, or they come from tropical savanna that is being settled rapidly by people. The tropical rain forests are the deep reservoirs of life on the planet, containing most of the world's plant and animal species. The rain forests are also its largest reservoirs of viruses, since all living things carry viruses. When viruses come out of an ecosystem, they tend to spread in waves through the human population, like echoes from the dying biosphere. Here are the names of some emerging viruses: Lassa. Rift Valley. Oropouche. Rocio. Q. Guanarito. VEE. Monkeypox. Dengue. Chikungunya. The hantaviruses. Machupo. Junin. The rabieslike strains Mokola and Duvenhage. LeDantec. The Kyasanur Forest brain virus. Then there is HIV—which is very much an emerging virus, because its penetration of the human species is increasing rapidly, with no end in sight. The Semliki Forest agent. Crimean-Congo. Sindbis. O'nyongnyong. Nameless São Paulo. Marburg. Ebola Sudan. Ebola Zaire. Ebola Reston.

In a sense, the earth is mounting an immune response against the human species. It is beginning to react to the human parasite, the flooding infection of people, the dead spots of concrete all over the planet, the cancerous rot-outs in Europe, Japan, and the United States, thick with replicating primates, the colonies enlarging and spreading and threatening to shock the biosphere with mass extinctions. Perhaps the biosphere does not "like" the idea of five bil-

lion humans. Or it could also be said that the extreme amplification of the human race, which has occurred only in the
past hundred years or so, has suddenly produced a very large
quantity of meat, which is sitting everywhere in the biosphere and may not be able to defend itself against a life
form that might want to consume it. Nature has interesting
ways of balancing itself. The rain forest has its own defenses. The earth's immune system, so to speak, has recognized the presence of the human species and is starting to
kick in. The earth is attempting to rid itself of an infection by
the human parasite. Perhaps AIDS is the first step in a natural
process of clearance.

AIDS is arguably the worst environmental disaster of the
twentieth century. The AIDS virus may well have jumped into
the human race from African primates, from monkeys and
anthropoid apes. For example, HIV-2 (one of the two major
strains of HIV) may be a mutant virus that jumped into us
from an African monkey known as the sooty mangabey,
perhaps when monkey hunters or trappers touched bloody
tissue. HIV-1 (the other strain) may have jumped into us from
chimpanzees—perhaps when hunters butchered chimpanzees. A strain of simian AIDS virus was recently isolated from
a chimpanzee in Gabon, in West Africa, which is, so far, the
closest thing to HIV-1 that anyone has yet found in the animal
kingdom.

The AIDS virus was first noticed in 1980 in Los Angeles by
a doctor who realized that his gay male patients were dying
of an infectious agent. If anyone at the time had suggested
that this unknown disease in gay men in southern California
came from wild chimpanzees in Africa, the medical community would have collectively burst out laughing. No one is
laughing now. I find it extremely interesting to consider the
idea that the chimpanzee is an endangered rain-forest animal
and then to contemplate the idea that a virus that moved
from chimps into us is suddenly not endangered at all. You

could say that rain-forest viruses are extremely good at looking after their own interests.

The AIDS virus is a fast mutator; it changes constantly. It is a hypermutant, a shape shifter, spontaneously altering its character as it moves through populations and through individuals. It mutates even in the course of one infection, and a person who dies of HIV is usually infected with multiple strains, which have all arisen spontaneously as mutants in the body. The fact that the virus mutates rapidly means that vaccines for it will be very difficult to develop. In a larger sense, it means that the AIDS virus is a natural survivor of changes in ecosystems. The AIDS virus and other emerging viruses are surviving the wreck of the tropical biosphere because they can mutate faster than any changes taking place in their ecosystems. They must be good at escaping trouble, if some of them have been around for as long as four billion years. I tend to think of rats leaving a ship.

I suspect that AIDS might not be Nature's preeminent display of power. Whether the human race can actually maintain a population of five billion or more without a crash with a hot virus remains an open question. Unanswered. The answer lies hidden in the labyrinth of tropical ecosystems. AIDS is the revenge of the rain forest. It may be only the beginning.

No problem, I thought. Of course, I'll be all right. We'll be all right. No problem at all. Everything will be all right. Plenty of people have gone inside Kitum Cave without becoming sick. Three to eighteen days. As the amplification begins, you feel nothing. It made me think of Joe McCormick, the C.D.C. official who had clashed with the Army over the management of the Ebola Reston outbreak. I remembered the story of him in Sudan, hunting Ebola virus. At the end of a plane flight into deep bush, he had come face to face with Ebola in a hut full of dying patients, had pricked his thumb with a bloody needle, and got lucky, and had survived the experience. In the end, Joe McCormick had been right about the Ebola Reston virus: it had not proved to be highly infectious in people. Then I thought about another Joe McCormick discovery, one of the few breakthroughs in the treatment of Ebola virus. In Sudan, thinking he was going to die of Ebola, he had discovered that a bottle of Scotch is the only good treatment for exposure to a filovirus.

I drove to the abandoned monkey house one day in autumn, to see what had become of it. It was a warm day in Indian summer. A brown haze hung over Washington. I turned off the Beltway and approached the building discreetly. The place was deserted and as quiet as a tomb. Out front, a sweet-gum tree dropped an occasional leaf. FOR LEASE signs sat in front of many of the offices around the parking lot. I sensed the presence not of a virus but of financial illness— clinical signs of the eighties, like your skin peeling off after a bad fever. I walked across the grassy area behind the building until I reached the Army's insertion point, a glass door. It was locked. Shreds of silver duct tape dangled from the door's edges. I looked inside and saw a floor mottled with reddish brown stains. A sign on the wall said CLEAN UP YOUR OWN MESS. Next to it, I discerned the air-lock corridor, the gray zone through which the soldiers had passed into the hot zone. It had gray cinder-block walls: the ideal gray zone.

My feet rustled through shreds of plastic in the grass. I found elderberries ripening around a rusted air-handling machine. I heard a ball bounce, and saw a boy dribbling a basketball on a playground. The ball cast rubbery echoes off the former monkey house. Children's shouts came from the day-care center through the trees. Exploring the back of the building, I came to a window and looked in. Climbing vines had grown up inside the room and had pressed against the glass of the windows, seeking warmth and light. Where had those vines found water inside the building? The vine was Tartarian honeysuckle, a weed that grows in waste places and on abandoned ground. The flowers of Tartarian honey-

suckle have no smell. That is, they smell like a virus; and
they flourish in ruined habitats. Tartarian honeysuckle
reminded me of Tartarus, the land of the dead in Virgil's
Aeneid, the underworld, where the shades of the dead whispered in the shadows.

I couldn't see through the tangled vines into the former
hot zone. It was like looking into a rain forest. I walked
around to the side of the building and found another glass
door beribboned with tape. I pressed my nose against the
glass and cupped my hands around my eyes to stop reflections, and saw a bucket smeared with a dry brown crust. The
crust looked like dried monkey excrement. Whatever it was,
I guessed it had been stirred up with Clorox bleach. A spider
had strung a web between a wall and the bucket of waste. On
the floor under the web, the spider had dropped husks of flies
and yellow jackets. The time of year being autumn, the spider had left egg cases in its web, preparing for its own cycle
of replication. Life had established itself in the monkey
house. Ebola had risen in these rooms, flashed its colors, fed,
and subsided into the forest. It will be back.

Main Characters

In order of appearance. (Military rank given as of the time of the Reston event.)

"Charles Monet." A French expatriate living in western Kenya. In January 1980, he essentially melts down with Marburg virus while traveling on an airplane.

Lieutenant Colonel Nancy Jaax. Veterinary pathologist at USAMRIID. Begins working with Ebola virus in 1983, when she gets a hole in her space-suit glove. Becomes chief of pathology at USAMRIID in 1989, and during the winter of that year becomes a player in the Reston biohazard operation.

Colonel Gerald ("Jerry") Jaax. Chief of the veterinary division at USAMRIID. Married to Nancy Jaax. Had never worn a biological space suit, but becomes the mission leader of the space-suited SWAT team during the Reston biohazard operation.

Ebola. (Pronounced ee-BOH-la.) Extremely lethal virus from the tropics, its exact origins unknown. It has three known subtypes: *Ebola Zaire, Ebola Sudan,* and *Ebola Reston.* It is closely related to **Marburg** virus. All of them constitute the **filovirus** family.

Eugene ("Gene") Johnson. Civilian virus hunter working for the Army. Specialist in Ebola. In the spring of 1988, following the death of "Peter Cardinal," leads an Army expedition to Kitum Cave in Mount Elgon. Chief of logistics and safety for the Reston biohazard operation.

"Peter Cardinal." Danish boy visiting his parents in Kenya in the summer of 1987, when he dies of Marburg virus. The Army keeps a strain of Marburg named after him in its freezers.

Dan Dalgard. Veterinarian at the Reston Primate Quarantine Unit (the Reston monkey house).

Peter Jahrling. Civilian Army virologist. Codiscoverer of the strain of virus that burns through the Reston monkey house.

Tom Geisbert. Intern at USAMRIID. In the fall of 1989, is responsible for the operation of USAMRIID's electron microscope. Codiscoverer of the virus.

Colonel Clarence James ("C. J.") Peters, MD. Chief of the disease-assessment division at USAMRIID. Overall leader of the Reston biohazard operation.

Major General Philip K. Russell, MD. The general who gave the command to dispatch the military teams to Reston.

Dr. Joseph B. McCormick. Chief of the Special Pathogens Branch of the C.D.C. Treated human Ebola patients in a hut in Sudan, where he stuck himself with a bloody needle.

Glossary

Amplification. Multiplication of a virus through either (1) the body of an individual host or (2) a population of hosts. See also **extreme amplification.**

Brick. (Military slang.) Pure crystal-like block of packed virus particles that grows inside a cell. Also known as an *inclusion body.* In this book, often called a *crystalloid* (author's own term).

Bubble stretcher. Portable biocontainment pod used for transportation of a hot patient.

Burn; burning. See **explosive chain of lethal transmission.**

Chemturion space suit. Pressurized, heavy-duty biological space suit used in Biosafety Level 4 containment areas. Also known as a *blue suit* because it is bright blue.

Crash and bleed out. (Military slang.) To die of shock, with profuse hemorrhages from the orifices of the body.

Crystalloid. See **brick.**

Decon. (Military slang.) To decontaminate; decontamination.

Electron microscope. Large and very powerful microscope that uses a beam of electrons to enlarge the image of a very small object, such as a virus, and replicate it on a screen.

Emerging viruses. "Viruses that have recently increased their incidence and appear likely to continue increasing." Term and definition coined by Stephen S. Morse, a virologist at Rockefeller University.

Envirochem. Green liquid disinfectant used in air-lock chemical showers. An effective virus killer.

Explosive chain of lethal transmission. Sort of biological meltdown wherein a lethal infectious agent spreads explosively through a population, killing a large percentage of the population. Also known as *burning*.

Extreme amplification. Multiplication of a virus everywhere in a host, partly transforming the host into virus.

Filovirus. A family of viruses that comprises only Ebola and Marburg. In this book, also called *thread viruses*.

Gray area; gray zone. Intermediate area or room between a hot zone and the normal world. A place where the two worlds meet.

Hatbox. (Military slang.) Cylindrical biohazard container made of waxed cardboard. Also known as an *ice-cream container*.

HIV. Human immunodeficiency virus, the cause of AIDS. It is an emerging Level 2 agent from the rain forests of Africa. Exact origin unknown. Now amplifying globally, its ultimate level of penetration into the human species is completely unknown. See also **amplification.**

Host. Organism that serves as a home to, and often as a food supply for, a parasite, such as a virus.

Hot. (Military slang.) Lethally infective in a biological sense.

Hot agent. Extremely lethal virus. Potentially airborne.

Hot suite. A group of Biosafety Level 4 laboratory rooms.

Hot zone; hot area; hot side. Area that contains lethal, infectious organisms.

Ice-cream container. See **hatbox.**

Index case. First known case in an outbreak of infectious disease. Sometimes spreads the disease widely.

Kinshasa Highway. AIDS highway. The main route by which HIV traveled during its breakout from the central African rain forest. The road links Kinshasa, in Zaire, with East Africa.

Marburg virus. Closely related to Ebola. Was initially called *stretched rabies.*

Mayinga strain. Hottest known strain of Ebola virus. Comes from a nurse known as Mayinga N., who died in Zaire in 1976.

Microbreak. (Author's own term.) Small, sometimes almost invisible outbreak of an emerging virus.

Nuke. (Military slang.) In biology, an attempt to render a place sterile. See also **sterilization.**

Racal suit. Portable, positive-pressure space suit with a battery-powered air supply. For use in fieldwork with extreme biohazards that are believed to be airborne. Also known as an *orange suit* because it is bright orange.

Replication. Self-directed copying. See also **amplification.**

Sentinel animal. Susceptible animal used as an alarm for the presence of a hot agent, since no instrument can detect a hot agent. Used as a canary is in a coal mine.

SHF. Simian hemorrhagic fever. A monkey virus that is harmless to humans.

Slammer. (Military slang.) The Biosafety Level 4 containment hospital at USAMRIID.

Sterilization. Unequivocal, total destruction of all living organisms. Extremely difficult to achieve in practice, and almost impossible to verify afterward.

Stretched rabies. See **Marburg virus.**

Submarine. (Military slang.) The Biosafety Level 4 morgue at USAMRIID.

Third spacing. Massive hemorrhagic bleeding under the skin.

USAMRIID. United States Army Medical Research Institute of Infectious Diseases, at Fort Detrick, in Frederick, Maryland. Also called *the Institute*.

Virus. Disease-causing agent smaller than a bacterium, consisting of a shell made of proteins and membranes and a core containing DNA or RNA. A virus depends on living cells in order to replicate.

Credits

I owe first thanks to the civilian and military staff of USAMRIID as a whole. The people who took part in the Reston operation risked their lives anonymously, with no expectation that their work would ever come to public attention.

I am deeply grateful to my editor at Random House, Sharon DeLano. At one point, I said to her, "God is in the details," and she replied, "No, God is in the structure." I am also indebted to Sally Gaminara, of Doubleday, U.K., for her valuable editorial ideas. Thanks for Ian Jackman's help, and also many thanks to Harold Evans. And to Charlie Conrad of Anchor Books.

For helping to keep my family solvent, much gratitude to Lynn Nesbit. Also: Robert Bookman, Lynda Obst, Cynthia Cannell, Eric Simonoff, and Chuck Hurewitz. Thanks to Jim Hart for his extremely perceptive conversations, and to Ridley Scott.

This book began as a *New Yorker* piece. I am grateful to Robert Gottlieb, who commissioned the article, and to Tina Brown, who published it and gave it wings. I am indebted to John Bennet, the editor for the piece, and to Caroline Fraser, the checker. Also many thanks to Pat Crow, Jill Frisch, Elizabeth Macklin, and Chip McGrath.

I received philosophical guidance from Stephen S. Morse

and Joshua Lederberg, both of whom are virologists at Rockefeller University in New York City. Some of the concerns (or fears) expressed in this book were brought to world attention in an important conference on emerging viruses organized and chaired by Morse, which happened, strangely, in May 1989, just months before the Reston outbreak. At that conference, Morse apparently coined the term "emerging virus." I have also been influenced by decades of thinking and commentary by Lederberg. Any scientific follies committed in this book are mine alone.

At USAMRIID, my special thanks go to Dr. Ernest Takafuji, commander of USAMRIID, and to David Franz, deputy commander. I also wish to acknowledge the detailed help of Peter Jahrling, Nancy and Jerry Jaax, Thomas Geisbert, and Eugene Johnson with passages that deal with their thoughts and feelings during the Reston crisis. Curtis Klages, Nicole Berke Klages, Rhonda Williams, and Charlotte Godwin Whitford also gave much time and help. Also thanks to: Cheryl Parrott, Carol Linden, Joan Geisbert, and Ed Wise, as well as to the other 91-Tangos and civilian animal caretakers who described their experiences at Reston to me. And many thanks to Ada Jaax.

At the Centers for Disease Control, for their generous time and for sharing their recollections: I thank Dr. C. J. Peters and Susan Peters, Dr. Joel Breman, Heinz Feldmann, Thomas G. Ksiazek, Dr. Joseph B. McCormick, and Anthony Sanchez. *With other institutions:* David Huxsoll, Dr. Frederick A. Murphy, and Dr. Philip K. Russell. *In Kenya:* Dr. Shem Musoke, Dr. David Silverstein, and Colonel Anthony Johnson. *In South Africa:* Dr. Margaretha Isaäcson and Dr. G. B. "Bennie" Miller. *On the Bighorn River:* Dr. Karl M. Johnson. *At Hazleton Washington:* I am grateful to Dan Dalgard for the assistance he gave me on portions of the manuscript that deal with his thoughts, as well as for letting me quote from his "Chronology of Events."

At the Alfred P. Sloan Foundation, I am most grateful to

Arthur L. Singer Jr., for his sustained interest and support. *At Princeton University,* thanks to Carol Rigolot, of the Council of the Humanities.

At Conservation International, thanks to Peter A. Seligmann and Russell Mittermeier. For the record, it was Mittermeier who seems to have originated the interesting comparison between the human species and a pile of meat waiting to be consumed.

Concerning the trip to Kitum Cave, I owe special thanks to Graham Boynton as well as to Christine Leonard, not to mention Robin and Carrie MacDonald, and Katana Chege, Morris Mulatya, Herman Andembe, and Jamy Buchanan. Ian Redmond gave me valuable information about the cave. Also, I cannot fail to mention the help of David and Gregory Chudnovsky.

Thanks to many friends: Peter Benchley, Freeman Dyson, Stona and Ann Fitch, Sallie Gouverneur, William L. Howarth, John McPhee, Dr. David G. Nathan, Richard O'Brien, Michael Robertson, Ann Waldron, Jonathan Weiner, and Robert H. White. Thanks to my grandfather, Jerome Preston, Sr., and my parents, Jerome Preston, Jr., and Dorothy Preston, for their support, and special thanks to my brother, Dr. David G. Preston, for his enthusiasm for the story, and to my other brother, the author Douglas Preston.

Final and greatest thanks go to my wife, Michelle Parham Preston, for her extraordinary support and love.